TRANSPORTATION AND THE ENVIRONMENT

Assessments and Sustainability

TRANSPORTATION AND THE ENVIRONMENT

Assessments and Sustainability

Edited by
Gabriela Ionescu, PhD

Apple Academic Press Inc. | Apple Academic Press Inc.
3333 Mistwell Crescent | 9 Spinnaker Way
Oakville, ON L6L 0A2 | Waretown, NJ 08758
Canada | USA

©2017 by Apple Academic Press, Inc.

First issued in paperback 2021

Exclusive worldwide distribution by CRC Press, a member of Taylor & Francis Group
No claim to original U.S. Government works

ISBN 13: 978-1-77-463697-8 (pbk)
ISBN 13: 978-1-77-188466-2 (hbk)

Library and Archives Canada Cataloguing in Publication

Transportation and the environment : assessments and sustainability/edited by Gabriela Ionescu, PhD.

Includes bibliographical references and index.
Issued in print and electronic formats.
ISBN 978-1-77188-466-2 (hardcover).--ISBN 978-1-77188-467-9 (pdf)
1. Transportation--Environmental aspects. 2. Roads--Environmental aspects. 3. Urban transportation--Environmental aspects. 4. Carbon dioxide mitigation. 5. Sustainable development. I. Ionescu, Gabriela, author, editor

HE147.65.T73 2016   388'.04   C2016-902423-7   C2016-902424-5

Library of Congress Cataloging-in-Publication Data

Names: Ionescu, Gabriela, editor.
Title: Transportation and the environment : assessments and sustainability / editor, Gabriela Ionescu, PhD.
Description: Toronto ; New Jersey : Apple Academic Press, [2016] | Includes bibliographical references and index.
Identifiers: LCCN 2016016119 (print) | LCCN 2016021677 (ebook) | ISBN 9781771884662 (hardcover : alk. paper) | ISBN 9781771884679 ()
Subjects: LCSH: Transportation--Environmental aspects. | Environmental impact analysis.
Classification: LCC TD195.T7 .T74525 2016 (print) | LCC TD195.T7 (ebook) | DDC 388.1/1--dc23
LC record available at https://lccn.loc.gov/2016016119

Apple Academic Press also publishes its books in a variety of electronic formats. Some content that appears in print may not be available in electronic format. For information about Apple Academic Press products, visit our website at **www.appleacademicpress.com** and the CRC Press website at **www.crc-press.com**

About the Editor

GABRIELA IONESCU, PhD

Dr. Gabriela Ionescu obtained her PhD in Power Engineering from Politehnica University of Bucharest and in Environmental Engineering from University of Trento. She is currently a member of the Department of Energy Production and Use at the Politehnica University of Bucharest and collaborator of the Department of Civil, Environmental and Mechanical Engineering at the University of Trento. She has done prolific research and has been published multiple times in areas related to energy efficiency, waste and wastewater management, energy conservation, life-cycle assessment, environmental analysis, and sustainability feasibility studies.

Contents

Acknowledgment and How to Cite

The editor and publisher thank each of the authors who contributed to this book. The chapters in this book were previously published in various places in various formats. To cite the work contained in this book and to view the individual permissions, please refer to the citation at the beginning of each chapter. Each chapter was read individually and carefully selected by the editor; the result is a book that provides a nuanced look at building sustainable transportation infrastructure. The chapters included are broken into three sections, which describe the following topics:

- Chapter 1 analyzes a few aspects of the role of a particular highway concerning the evolution of its importance at the international level, the difference in terms of vehicle park, when compared with the local park of the crossed territories, the relevance of the engine evolution in the transport sector, in order to understand the importance of updating a European modeling approach.
- Chapter 2 identifies specific local roadway infrastructure design guidelines associated with the construction and operation of sustainable energy source facilities, such as ethanol plants, biomass plants, and wind farm facilities.
- Chapter 3 examines the management of pavement network roughness, using a life-cycle approach to assess changes in total greenhouse gas (measured in carbon dioxide emissions) from strategic management of highway pavement roughness.
- Chapter 4 uses life-cycle impact assessments of the new bus rapid transit and light rail lines in Los Angeles.
- Chapter 5 assessed environmental impacts from 181 planned transportation projects in the San Francisco Bay Area, with findings that apply well to other metropolitan areas.
- Chapter 6 gives a framework for quantifying the burdens of ground transportation in urban settings that incorporates travel time, vehicle fuel, and pavement maintenance costs.
- Chapter 7 contains an initial review of the issues surrounding sustainable development and airport surface access, focusing on two aspects: an evaluation of the technological innovation options that will enable sustainable transport solutions for surface access trips, and a discussion of the role of behavioral change for these journeys.

- Chapter 8 discusses two different types of pure biodiesel fuel, namely, methanol-based biodiesel and ethanol-based biodiesel and found that ethanol-based biodiesel blends result in higher smoke emissions than pure diesel fuel, while methanol-based biodiesel blends smoke emissions are lower compared to pure diesel fuel.
- Chapter 9 confirms that the electric car can serve as a suitable instrument toward a much more sustainable future in mobility.
- Chapter 10 confirms that bio-jet fuel produced from non-edible oilseeds can be an alternative to fossil fuels, with the added benefits of increasing national energy security, reducing environmental impact, and fostering rural economic growth.
- Chapter 11 summarizes the reasons why it is imperative that we change how we harness and use energy for transport.

List of Contributors

Karim A. Abdel Warith
School of Civil Engineering, Purdue University: Indiana Local Technical Assistance Program

Imad Ahmed
School of Electronic and Electrical Engineering, University of Leeds, Leeds, UK

Panagiotis Ch. Anastasopoulos
Department of Civil, Structural and Environmental Engineering, Institute for Sustainable Transportation and Logistics, University at Buffalo, The State University of New York

Tom Budd
Transport Studies Group, School of Civil & Building Engineering, Loughborough University; Loughborough, UK

Charalambos A. Chasos
Department of Mechanical Engineering, Frederick University, 1036 Nicosia, Cyprus

Mikhail Chester
Civil, Environmental, and Sustainability Engineering, School of Sustainability, Arizona State University, 501 E Tyler Mall Room 252, Mail Code 5306, Tempe, AZ 86287-5306, USA

Chris N. Christodoulou
Department of Mechanical Engineering, Frederick University, 1036 Nicosia, Cyprus

Andres F. Clarens
Department of Civil and Environmental Engineering, University of Virginia, Thornton Hall B228, 351 McCormick Road, PO Box 400742, Charlottesville, VA 22904-4742, USA

William Eisenstein
Center for Resource Efficient Communities, University of California, Berkeley, 390 Wurster Hall #1839, Berkeley, CA 94720, USA

Zoe Elizabeth
Institute of the Environment and Sustainability, California Center for Sustainable Communities, University of California, Los Angeles, LaKretz Hall, Suite 300, 619 Charles E Young Dr. East, Los Angeles, CA 90095-1496, USA

Jaafar Elmirghani
School of Electronic and Electrical Engineering, University of Leeds, Leeds, UK and Department of Electrical and Computer Engineering, King Abdulaziz University, Jeddah, Kingdom of Saudi Arabia

Jon D. Fricker
School of Civil Engineering, Purdue University: Indiana Local Technical Assistance Program

Conrad A. Gosse
Department of Civil and Environmental Engineering, University of Virginia, Thornton Hall B228, 351 McCormick Road, PO Box 400742, Charlottesville, VA 22904-4742, USA

John E. Haddock
School of Civil Engineering, Purdue University: Indiana Local Technical Assistance Program

John Harvey
University of California Pavement Research Center (UCPRC, Davis), 3153 Ghausi Hall, One Shields Avenue, Davis, CA 95616, USA

Eckard Helmers
Institut für angewandtes Stoffstrommanagement (IfaS) am Umwelt-Campus Birkenfeld, Trier University of Applied Sciences, P.O. Box 1380, Birkenfeld, D-55761, Germany

Yinbin Huang
Department of Agricultural and Biosystems Engineering, South Dakota State University, Brookings, USA

Patrick R. Huber
Department of Environmental Science and Policy, University of California, Davis, California, USA

Gabriela Ionescu
University of Trento, Civil Environmental and Mechanical Engineering Department, via Mesiano 77, 38123, Trento, Italy

James Julson
Department of Agricultural and Biosystems Engineering, South Dakota State University, Brookings, USA

George N. Karagiorgis
Department of Mechanical Engineering, Frederick University, 1036 Nicosia, Cyprus

Alissa Kendall
Department of Civil and Environmental Engineering, UC Davis, 3167 Ghausi Hall, One Shields Avenue, Davis, CA 95616, USA

Patrick Marx
Institut für angewandtes Stoffstrommanagement (IfaS) am Umwelt-Campus Birkenfeld, Trier University of Applied Sciences, P.O. Box 1380, Birkenfeld, D-55761, Germany

Keith Mason
Department of Air Transport, School of Engineering, Cranfield University, Cranfield, UK

Juan Matute
Luskin School of Public Affairs, Local Climate Change Initiative, University of California, Los Angeles, Box 165606, 3250 Public Affairs Building, Los Angeles, CA 90095-1656, USA

Chikage Miyoshi
Department of Air Transport, School of Engineering, Cranfield University, Cranfield, UK

Richard Moxon
Department of Air Transport, School of Engineering, Cranfield University, Cranfield, UK

Elizabeth O'Donoghue
The Nature Conservancy, 201 Mission Street, 4th Floor, San Francisco, CA 94105, USA

Valeriu Nicolae Panaitescu
University Politehnica of Bucharest, Power Engineering Faculty, Department of Hydraulic, Hydraulical Machinery and Environment Engineering, 313 Splaiul Independentei, 060042, Bucharest, Romania

Stephanie Pincetl
Institute of the Environment and Sustainability, California Center for Sustainable Communities, University of California, Los Angeles, LaKretz Hall, Suite 300, 619 Charles E Young Dr. East, Los Angeles, CA 90095-1496, USA

Bilal Qazi
School of Electronic and Electrical Engineering, University of Leeds, Leeds, UK

Elena Cristina Rada
University of Trento, Civil Environmental and Mechanical Engineering Department, via Mesiano 77, 38123, Trento, Italy

Marco Ragazzi
University of Trento, Civil Environmental and Mechanical Engineering Department, via Mesiano 77, 38123, Trento, Italy

Wayne Richardson
Bertsch-Frank & Associates, LLC

Tim Ryley
Transport Studies Group, School of Civil & Building Engineering, Loughborough University; Loughborough, UK

Maria J. Santos
Department of Environmental Sciences, Copernicus Institute of Sustainable Development, Utrecht University, Heidelberglaan 2, 3584 CS Utrecht, Netherlands

Daniel Sperling
Professor of Civil Engineering and Environmental Science and Policy, and Director of the Institute of Transportation Studies, University of California, Davis

Constantin-Cristian Stroe
University Politehnica of Bucharest, Power Engineering Faculty, Department of Hydraulic, Hydraulical Machinery and Environment Engineering, 313 Splaiul Independentei, 060042, Bucharest, Romania

James H. Thorne
Department of Environmental Science and Policy, University of California, Davis, California, USA

Ting Wang
University of California Pavement Research Center (UCPRC, Davis), 2001 Ghausi Hall, One Shields Avenue, Davis, CA 95616, USA

Lin Wei
Department of Agricultural and Biosystems Engineering, South Dakota State University, Brookings, USA

Alberto Zanni
Transport Studies Group, School of Civil & Building Engineering, Loughborough University; Loughborough, UK

Xianhui Zhao
Department of Agricultural and Biosystems Engineering, South Dakota State University, Brookings, USA

Introduction

The environmental impact of transportation is significant. Transportation is a major user of energy, and it burns most of the world's petroleum. This creates air pollution, including nitrous oxides and particulates, and is a significant contributor to global warming through carbon dioxide emissions. In fact, transportation is the fastest-growing contributor to carbon dioxide emissions.

Although environmental regulations in many countries have reduced the individual vehicle's emissions, this has been offset by an increase in the number of vehicles on the road and airways. Methods to reduce the carbon emissions of vehicles need intensive ongoing research. New means of transportation must be developed alongside greater energy efficiency. By globally reducing transportation emissions, Earth's air quality will be improved; acid rain and smog will be reduced; and climate change effects will be reduced.

This means that the research collected here in this book is of vital importance for today and for tomorrow. It forms the foundation for a new approach to transportation.

—Gabriela Ionescu

In Chapter 1, Stroe and colleagues analyse a few aspects of the role of the Italian A22 highway (the Brenner highway) concerning the evolution of its importance at international level, the difference in terms of vehicle park, when compared with the local park of the crossed territories, the relevance of the engine evolution in the transport sector. Data from the iMonitraf European project and from the Italian ACI database are used. The above analysis is aimed to understand the importance of updating the modelling approach used in the ALPNAP European project.

Chapter 2, by Abdel Warith and colleagues, aims to identify specific local roadway infrastructure design guidelines associated with the construction and operation of sustainable energy source facilities, such as ethanol plants, biomass plants, and wind farm facilities. Data associated with sustainable energy facility traffic in Indiana were collected to develop Excel-based tools (worksheets) and assist local agencies in the design of pavements in the proximity of ethanol plants, biomass plants, and wind farms. To that end, a simple procedure is presented, which provides a design capable of withstanding heavy traffic loads, while, at the same time, quantifies the effects that new sustainable energy source facilities may have on local road networks. The procedure is accompanied by two MS Excel-based software tools that can be used in the design of local roads adjacent to such sustainable energy facilities. The developed worksheets can serve as a hands-on tool to assist local government engineers in evaluating and in quantifying the probable effects of the construction and operation of a sustainable energy facility in their jurisdiction.

On-road vehicle use is responsible for about a quarter of US annual greenhouse gas (GHG) emissions. Changes in vehicles, travel behavior and fuel are likely required to meet long-term climate change mitigation goals, but may require a long time horizon to deploy. Chapter 3, by Wang and colleagues, examines a near-term opportunity: management of pavement network roughness. Maintenance and rehabilitation treatments can make pavements smoother and reduce vehicle rolling resistance. However, these treatments require material production and equipment operation, thus requiring a life cycle perspective for benefits analysis. They must also be considered in terms of their cost-effectiveness in comparison with other alternatives for affecting climate change. The chapter describes a life cycle approach to assess changes in total GHG (measured in CO_2-e) emissions from strategic management of highway pavement roughness. Roughness values for triggering treatments are developed to minimize GHG considering both treatment and use phase vehicle emission. With optimal triggering for GHG minimization, annualized reductions on the California state highway network over a 10-year analysis period are calculated to be 0.82, 0.57 and 1.38 million metric tons compared with historical trigger values, recently implemented values and no strategic intervention (reactive maintenance), respectively. Abatement costs calculated using $/metric-ton

CO_2-e are higher than those reported for other transportation sector abatement measures, however, without considering all benefits associated with pavement smoothness, such as vehicle life and maintenance, or the time needed for deployment.

Public transportation systems are often part of strategies to reduce urban environmental impacts from passenger transportation, yet comprehensive energy and environmental life-cycle measures, including upfront infrastructure effects and indirect and supply chain processes, are rarely considered. Using the new bus rapid transit and light rail lines in Los Angeles, in Chapter 4 Chester and colleagues develop near-term and long-term life-cycle impact assessments, including consideration of reduced automobile travel. Energy consumption and emissions of greenhouse gases and criteria pollutants are assessed, as well the potential for smog and respiratory impacts. Results show that life-cycle infrastructure, vehicle, and energy production components significantly increase the footprint of each mode (by 48–100% for energy and greenhouse gases, and up to 6200% for environmental impacts), and emerging technologies and renewable electricity standards will significantly reduce impacts. Life-cycle results are identified as either local (in Los Angeles) or remote, and show how the decision to build and operate a transit system in a city produces environmental impacts far outside of geopolitical boundaries. Ensuring shifts of between 20–30% of transit riders from automobiles will result in passenger transportation greenhouse gas reductions for the city, and the larger the shift, the quicker the payback, which should be considered for time-specific environmental goals.

Globally, urban areas are expanding, and their regional, spatially cumulative, environmental impacts from transportation projects are not typically assessed. However, incorporation of a Regional Advance Mitigation Planning (RAMP) framework can promote more effective, ecologically sound, and less expensive environmental mitigation. As a demonstration of the first phase of the RAMP framework, in Chapter 5 Thorne and colleagues assessed environmental impacts from 181 planned transportation projects in the 19 368 km^2 San Francisco Bay Area. They found that 107 road and railroad projects will impact 2411–3490 ha of habitat supporting 30–43 threatened or endangered species. In addition, 1175 ha of impacts to agriculture and native vegetation are expected, as well as 125 crossings of waterways supporting anadromous fish species. The extent of these spatially cumulative impacts shows the need for a regional approach to associated environmental offsets. Many of the impacts were comprised of numerous small projects, where project-by-project mitigation would result in increased transaction costs, land costs, and lost project time. Ecological gains can be made if a regional approach is taken through the avoidance of small-sized reserves and the ability to target parcels for acquisition that fit within conservation planning designs. The methods are straightforward, and can be used in other metropolitan areas.

Efforts to reduce the environmental impacts of transportation infrastructure have generally overlooked many of the efficiencies that can be obtained by considering the relevant engineering and economic aspects as a system. In Chapter 6, Gosse and Clarens present a framework for quantifying the burdens of ground transportation in urban settings that incorporates travel time, vehicle fuel and pavement maintenance costs. A Pareto set of bi-directional lane configurations for two-lane roadways yields non-dominated combinations of lane width, bicycle lanes and curb parking. Probabilistic analysis and microsimulation both show dramatic mobility reductions on road segments of insufficient width for heavy vehicles to pass bicycles without encroaching on oncoming traffic. This delay is positively correlated with uphill grades and increasing traffic volumes and inversely proportional to total pavement width. The response is nonlinear with grade and yields mixed uphill/downhill optimal lane configurations. Increasing bicycle mode share is negatively correlated with total costs and emissions for lane configurations allowing motor vehicles to safely pass bicycles, while the opposite is true for configurations that fail to facilitate passing. Spatial impacts on mobility also dictate that curb parking exhibits significant spatial opportunity costs related to the total cost Pareto curve. The proposed framework provides a means to evaluate relatively inexpensive lane reconfiguration options in response to changing modal share and priorities. These results provide quantitative evidence that efforts to reallocate limited pavement space to bicycles, like those being adopted in several US cities, could appreciably reduce costs for all users.

Sustainable development reflects an underlying tension to achieve economic growth whilst addressing environmental challenges, and this is particularly the case for the aviation sector. Although much of the aviation-related focus has fallen on reducing aircraft emissions, airports have also been under increasing pressure to support the vision of a low carbon energy future. One of the main sources of airport-related emissions is passenger journeys to and from airports (the surface access component of air travel), which is the focus of Chapter 7, by Ryley and colleagues. Two aspects associated with the relationship between sustainable development and airport surface access are considered. Firstly, there is an evaluation of three technological innovation options that will enable sustainable transport solutions for surface access journeys: telepresence systems to reduce drop-off/pick-up trips, techniques to improve public transport and options to encourage the sharing of rides. Secondly, the role of behavioral change for surface access journeys from a theoretical perspective, using empirical data from Manchester airport, is evaluated. Finally, the contribution of technology and behavioral intervention measures to improvements in sustainable development are discussed.

There is a recent interest for the utilisation of renewable and alternative fuel, which is regulated by the European Union, that currently imposes a lower limit of 7% by volume of biodiesel fuel blend in diesel fuel. The biodiesel physical characteristics, as well as the percentage of biodiesel blend in diesel fuel, affect the injector nozzle flow, the spray characteristics, the resulting air/fuel mixture, and subsequently the combustion quality and emissions, as well as the overall engine performance. In Chapter 8, bu Chasos and colleagues, two different types of pure biodiesel fuel, namely, methanol-based biodiesel and ethanol-based biodiesel, were produced in the laboratory of Frederick University by chemical processing of raw materials. The two biodiesel fuels were used for blending pure diesel fuel at various percentages. The blends were used for smoke emissions measurements of a diesel internal combustion engine at increasing engine speed and for increasing engine temperatures. From the experimental investigations it was found that ethanol-based biodiesel blends result in higher smoke emissions than pure diesel fuel, while methanol-based biodiesel blends smoke emissions are lower compared to pure diesel fuel.

Electric vehicles have been identified as being a key technology in reducing future emissions and energy consumption in the mobility sector. The focus of Chapter 9, by Helmers and Marx, is to review and assess the energy efficiency and the environmental impact of battery electric cars

(BEV), which is the only technical alternative on the market available today to vehicles with internal combustion engine (ICEV). Electricity on-board a car can be provided either by a battery or a fuel cell (FCV). The technical structure of BEV is described, clarifying that it is relatively simple compared to ICEV. Following that, ICEV can be 'e-converted' by experienced personnel. Such an e-conversion project generated reality-close data reported here. Practicability of today's BEV is discussed, revealing that particularly small-size BEVs are useful. This article reports on an e-conversion of a used Smart. Measurements on this car, prior and after conversion, confirmed a fourfold energy efficiency advantage of BEV over ICEV, as supposed in literature. Preliminary energy efficiency data of FCV are reviewed being only slightly lower compared to BEV. However, well-to-wheel efficiency suffers from 47% to 63% energy loss during hydrogen production. With respect to energy efficiency, BEVs are found to represent the only alternative to ICEV. This, however, is only true if the electricity is provided by very efficient power plants or better by renewable energy production. Literature data on energy consumption and greenhouse gas (GHG) emission by ICEV compared to BEV suffer from a 25% underestimation of ICEV-standardized driving cycle numbers in relation to street conditions so far. Literature data available for BEV, on the other hand, were mostly modeled and based on relatively heavy BEV as well as driving conditions, which do not represent the most useful field of BEV operation. Literature data have been compared with measurements based on the converted Smart, revealing a distinct GHG emissions advantage due to the German electricity net conditions, which can be considerably extended by charging electricity from renewable sources. Life cycle carbon footprint of BEV is reviewed based on literature data with emphasis on lithium-ion batteries. Battery life cycle assessment (LCA) data available in literature, so far, vary significantly by a factor of up to 5.6 depending on LCA methodology approach, but also with respect to the battery chemistry. Carbon footprint over 100,000 km calculated for the converted 10-year-old Smart exhibits a possible reduction of over 80% in comparison to the Smart with internal combustion engine. Findings of the article confirm that the electric car can serve as a suitable instrument towards a much more sustainable future in mobility. This is particularly true for small-size BEV, which is underrepresented in LCA literature data so far. While CO_2-LCA of BEV seems to be relatively well known apart from the battery, life cycle impact of BEV in categories other than the global warming potential reveals a complex and still incomplete picture. Since technology of the electric car is of limited complexity with the exception of the battery, used cars can also be converted from combustion to electric. This way, it seems possible to reduce CO_2-equivalent emissions by 80% (factor 5 efficiency improvement).

Bio-jet fuel produced from non-edible oilseeds can be an alternative to fossil fuels with the benefits of increasing national energy security, reducing environmental impact, and fostering rural economic growth. Efficient oil extraction from oilseeds is critical for economic production of bio-jet fuels. In Chapter 10, Zhao and colleagues conducted oil extractions from camelina (sativa) and canola (*Brassica napus*) seeds using a cold press method. The effect of the frequency controlling the screw rotation speed on the oil recovery and quality was discussed. Characterization of the produced raw vegetable oils, such as heating value, elemental content and main chemical compositions, was carried out. The results showed that the oil recovery increased when the frequency decreased. The highest oil recoveries for camelina and canola seeds were 88.2% and 84.1% respectively, both at 15 Hz. The cold press frequency and processing temperature (97.2°C - 106.0°C) had a minor influence on the qualities and recovery of both camelina and canola oils. In addition, camelina and canola oils produced at 15 Hz underwent catalytic cracking to examine potential hydrocarbon fuels production. It was observed that some of oil physicochemical properties were improved after catalytic cracking. Although more study is needed for further improvement of oil recovery and qualities, cold press could be an efficient method for oil extraction from non-edible oilseeds. Additionally, the preliminary results of upgrading the oils produced show very promising for future bio-jet fuels production.

In Chapter 11, Sperling argues that we need to wean ourselves off fossil fuels—and the GHG emissions they produce—and rebuild our cities and transportation systems to be far more energy efficient. We need to shift to a world that relies on sun, water, wind, and plants for energy. We need visions, strategies, and action.

PART I

ROADWAYS

CHAPTER 1

Some Considerations on the Environmental Impact of Highway Traffic

CONSTANTIN-CRISTIAN STROE,
VALERIU NICOLAE PANAITESCU, MARCO RAGAZZI,
ELENA CRISTINA RADA, AND GABRIELA IONESCU

Road transport emissions have straight forward effects on air quality requirements, at local, regional and even global scales [1-3]. Transport is responsible for around a quarter of EU greenhouse gas emissions making it the second biggest greenhouse gas-emitting sector after energy generation [4]. Due to the vehicles engine internal combustion and fossil fuels usage, transport has become a significant source of air pollutants including CO, CO_2, volatile organic compounds, NOx, NO_2, O_3, total suspended particles and particulate matter (PM). Several epidemiological investigations have studied the toxicity of certain pollutants resulted from traffic emission with acute and chronic health effects [5-7]. Results cannot be generalized as the impact of a highway depends on vehicle characteristics and number, local climatology and use of land.

Some Considerations on the Environmental Impact of Highway Traffic. © Stroe C-C, Panaitescu VN, Ragazzi M, Rada EC, and Ionescu G. Revista de Chimie 65,2 (2014), http://www.revistadechimie.ro/ pdf/STROE%20C.pdf%202%2014.pdf. Reprinted with permission from the authors and the publisher.

In particular, emission rates depend on the traffic flow, fleet, speed, vehicles characteristics such as type, size, age of a vehicle, engine condition, exhaust emission control systems, vehicle maintenance and weight [8,9].

Moreover, to all these, local and synoptic-scale meteorological conditions, season, topography and atmospheric chemical processes can be added. The transboundary long-range transport emission can be a source of episodic pollution caused by transnational pollutants. In comparison with flat terrain, in the mountain regions, the high-pressure conditions can produce the stagnation and recirculation of the emissions in the valleys and basin to which it may be added the vertical stability that can reduce the air pollution dispersion in the Alps [10,11].

The Brenner highway (A22) is the most important traffic artery connecting Italy with the rest of Europe by crossing the Alps with an altitude of 1350 m above sea level at the pass and covering almost 312 km.

The EU legislation imposes air quality limitations through Directive 2008/50/EC (to be met before 2015) expressed in pollutant mass concentrations by volume correlated with the exposure time and depending on the kind of compound. To this concern, in order to study the role of the A22 highway in terms of local impact on air quality, two studies were developed in the last decade: ALPNAP (The Air Pollution, Traffic Noise and Related Health Effects in the Alpine Space project) and iMonitraf (Monitoring of Road Traffic related Effects in Alpine Space and Common Measures) [12,13].

The present paper starts from the last quoted study in order to analyze and compare some aspects of the environmental impact of A22 with other important European arteries. The adopted approach can be a starting point for future explorations and proposals for viable solutions for emissions reduction.

1.1 EXPERIMENTAL PART

1.1.1 MATERIALS AND METHODS

For the development of the present paper, three pollutants (CO, NOx, PM_{10}) and five highways were considered, as shown in figure 1.

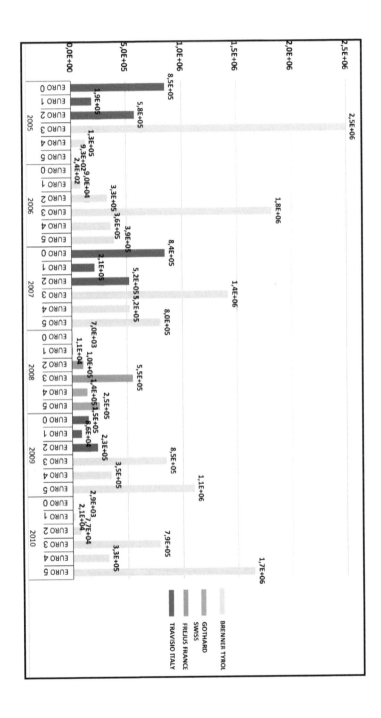

FIGURE 2: Maximum number of vehicles

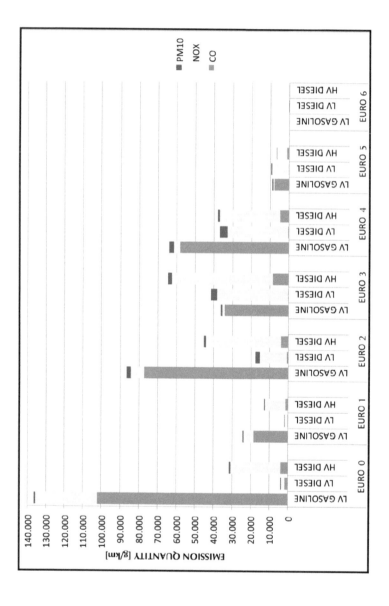

FIGURE 3: Total amount of emissions (per km) according to vehicle type, fuel type and Euro class (2010) in the Province of Trento

The Fréjus Road Tunnel is a tunnel that connects France and Italy being one of the major trans-Alpine transport routes. It is used for 80% of the commercial road traffic between the two countries.

The Mont Blanc Tunnel is a road tunnel in the Alps under the Mont Blanc mountain, linking France and Italy. One-third of Italian freight to northern Europe passes through this tunnel.

The St. Gotthard Tunnel forms part of the Swiss A2 motorway, running south from Basel through the tunnel down to Chiasso on the border with Italy.

The Tarvisian region is located along one of the main European thoroughfares: the Wien-Venice-Rome line. The main road that brings to Tarvisio is the A23 Alpe Adria highway.

Brenner is a small municipality in South Tyrol in northern Italy. Brenner lies about 110 km northeast of the city of Trento and about 60 km north of the city of Bolzano, on the border with Austria. The south end of the Brenner Pass is located in this municipality. The highway on the Italian side is named A22.

Statistical data regarding the vehicles that cross the five considered areas were extracted from iMonitraf and from an Italian database, that takes into account the Euro class and the fuel types of the vehicles park at regional level [14]. Three analyses were performed:

- relative importance of the A22 at international level taking into account the above mentioned highway and the characteristics of their vehicle streams;
- differences between the vehicle park of A22 and the one of one of the crossed regions;
- crossed significance of Euro classes, depending on fuel, weight of vehicle, pollutant.

After these preliminary analyses some considerations on the importance of updating of the ALPNAP diffusion modelling are presented.

1.2 RESULTS AND DISCUSSIONS

In the present work, the above-cited highways located in mountain regions were analyzed in order to give a synthetic vision of their environmental impact. Due to the relevance of the data, only 4 of them are presented in figure 2 concerning the maximum number of vehicles for Euro class cross-ing the considered highways, year by year.

From 2005 to 2010, it can be observed that the relative importance of Brenner highway is increasing. In particular, in 2010 the A22 pass results the more important one even if it registers a contraction in the passages. Indeed, according to iMonitraf the yearly number of passages through the Brenner pass decreased from about 3.3 millions of 2005 to less than 2.9 millions in 2010. The recent fluctuation can be attributed both to an evolution of the international organisation of transports (also by train) and on the recent economic crisis. Moreover, from figure 2 it can be seen that A22 resulted the most important as Euro 3 passages in 2005, but as Euro 5 in 2010.

In figure 3 the total amount of pollutants emitted from 1 km drive for each kind of vehicle is reported for the province of Bolzano, crossed by A22. It is clear that the engine evolution significantly decreases the emis-sions [15].

In term of mass comparison, as expected, the importance of diesel heavy duty vehicles is clear, but not dominant as data concern the overall province. Zooming on the sector of heavy duty vehicles, a comparison be-tween the park in the province and the park passing through the highway can be made, as shown in figures 4 and 5. It can be noticed that the trucks on the A22 have younger engines compared to the ones on the province. That can be explained by the fact that the trucks used for long distances need to guarantee good reliability.

The presence of better vehicular park cannot be seen as a guarantee of moderate impact on the territory as the emissions of the A22 are linear.

The assessment of the environmental impact of a highway is a com-plex problem, as demonstrated from the articulated significance of the Euro class, whose emission relevance changes from vehicle to vehicle de-pending on fuel and dimensions. To this concern, an analysis of the current situation that show how many light vehicles can produce the same amount of pollutant for one heavy duty vehicle is presented in figures 6, 7, and 8. It is clear that the evolution of the engines from Euro 0 to Euro 5 gave sig-nificant results. Euro 6 is expected to contribute to the reduction of NOx.

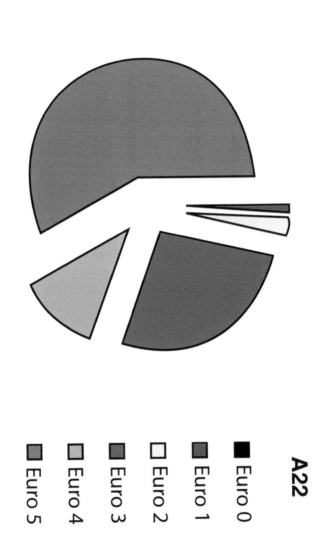

FIGURE 4: Heavy duty diesel vehicle for A22 highway (2010)

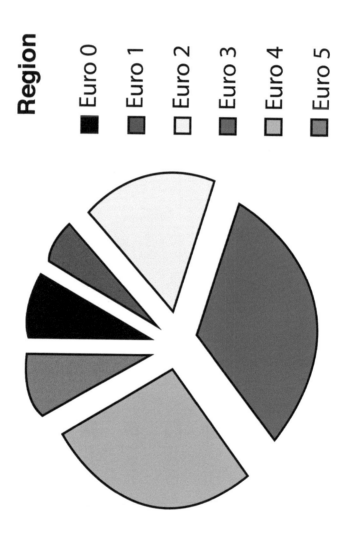

FIGURE 5: Heavy duty diesel vehicle for the Bolzano Province (2010)

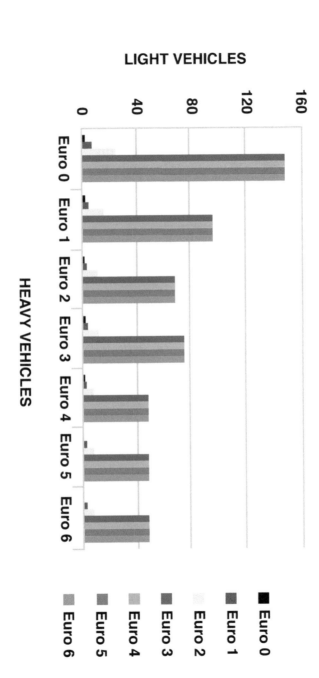

FIGURE 6: Heavy per light vehicle ratio for CO emissions

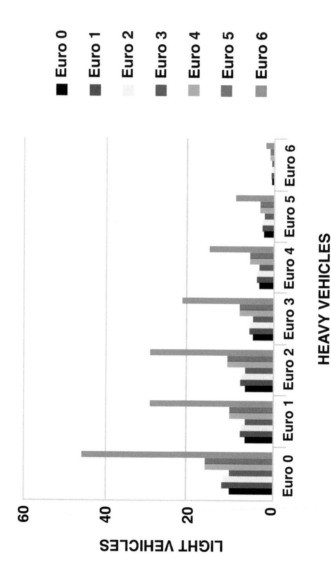

FIGURE 7: Heavy per light vehicle ratio for NOx emissions

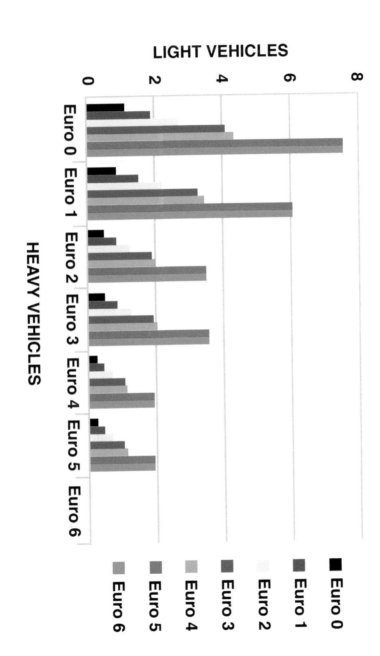

FIGURE 8: Heavy per light vehicle ratio for PM10 emissions

Taking into account the above analysis, the following considerations can be made:

- the traffic on the A22 highway is significant, but is related to an important trend in terms of engines quality;
- the introduction of Euro 6 will contribute to the solution of a lot of problems;
- A22 highway, for which is the incidence on the yearly average of NOx concentration along the road;
- the engine distribution at regional level is not proportional to their impact as the number of km driven changes from class to class; on the contrary, the engine percentage of a highway is exactly what is measured during emission.

1.3 CONCLUSIONS

In the last years, freight traffic through Alps is growing significantly, particularly the road transport, resulting a potential increased pollution, congestion, and safety concerns if the evolution of the environmental performances of transport does not counterbalance this dynamics.

It is clear that the assessment of the environmental impact of a highway like the A22 needs both historic data on vehicles, to generate trend for the future, and meteorological data on the territor y, as input of meteorological pre-processors for diffusion models. The project ALPNAP demonstrated that the quantity and quality of such data are available for the A22 case. However this project should be updated, taking into account that recent vehicular data can be obtained by the iMonitraf project. The ALPNAP updating has been planned by the authors in order to understand if the expected evolution of the traffic and of the engines can face with the problem of local impact in terms of NOx incidence.

The decision-makers try to improve their knowledge by a better comprehension of the link between environmental, social and economic dimension and the interdependence between territorial and temporal scales. This knowledge supposes a global approach, which has to be able to give elements for estimation, measurement and follow-up.

REFFERENCES

1. Rada, E.C., Ragazzi, M., Brini, M., Marmo, L., Zambelli, P., Chelodi, M., Ciolli, M., Sci. Bull., series D, 74, nr. 2, 2012, p. 2432.
2. Torretta, V., Rada, E.c., Panaitescu, V., Apostol, T. , Sci. Bull., series D, 74, 2012, p. 241.
3. Ionescu, G., Apostol, T., Rada, E.C., Ragazzi, M., Torretta, V., Sci. Bull., series D, 75, nr. 2, 2013, p. 25.
4. EEA-European Environment Agency, Climate Policies on Transport, http://ec.europa.eu, 2011.
5. Butu, I. M., Butu M., Constantinescu I., Macovei S., Rev. Chim. (Bucharest), 63, no. 3, 2012, p. 330.
6. Ionescu G., Zardi D., Tirler W., Rada E.C., Ragazzi M., Sci. Bull., series D, 74, nr. 4, 2012, p. 227.
7. Zhang, K., Batterman, S., Sci. Total Environ., 450-451, 2013, p. 307.
8. Buonanno G., Lall A.A., Stabile L., Atmos. Environ., 43, 2009, p.1100.
9. Pandian S., Gokhale S., Ghoshal A. K., Transport. Res., Part D, 14, 2009, p. 180.
10. Serafin, s., Zardi, d., J. Atmos. Sci., 67, nr. 4, 2010, p. 1171.
11. Ciuta, S., Schiavon, M., Chistè, A., Ragazzi, M., Rada, E.C., Tubino, M., Badea, A., Apostol, T., Sci. Bull., 74, nr. 1, series D, 2012, p. 211.
12. Heimann, D., Clemente, M., Elampe, E., Only, X., Miège, B., Defrance, J., Baulac, M., Suppan, P., Schäfer, K., Emeis, S., Forkel, R., Trini Castelli, S., Anfossi, D., Belfiore, G., Lercher, P., Rüdisser, J., Uhrner, U., Öttl, D., Rexeis, M., De Franceschi, M., Zardi, Cocarta, D., Ragazzi, M., Antonacci, G., Cemin, A., Seibert, P., Schicker, I., Krüger, B., Obleitner, F., Vergeiner, J., Grießer, E., Botteldooren, D., Renterghem, T. Van, "Air Pollution, Traffic Noise and Related Health Effects in the Alpine Space", University of Trento, DICA, 2007, p. 335.
13. iMonitraf, 2012, www.imonitraf.org.
14. ACI, 2011, www.aci.it.
15. Negoitescu, A., Ostoia, D., Tokar, A., Hamat, C., Rev. Chim.(Bucharest), 60, no. 4, 2009, p. 411

Figure 1 is not available in this version of the article. To view this additional information, please use the citation on the first page of this chapter.

CHAPTER 2

Design of Local Roadway Infrastructure to Service Sustainable Energy Facilities

KARIM A. ABDEL WARITH, PANAGIOTIS CH. ANASTASOPOULOS, WAYNE RICHARDSON, JON D, FRICKER, AND JOHN E. HADDOCK

2.1 BACKGROUND

Renewable, sustainable energy sources are being developed at a record pace throughout the USA and globally, with multidimensional benefits, as they have the potential to boost local economies and generate new jobs [1–20]. In Indiana, energy corporations have invested in three main types of sustainable energy sources, namely, ethanol, wind, and biomass energy, and have built numerous wind farms and ethanol and biomass plants. It is expected that the number of plants and wind farms will triple by 2022 [21]. Increased loads, increased traffic, or both can negatively affect road networks (with respect to the existing infrastructures, the environment, the aesthetics of the local communities, and the safety of the neighboring residents) when sustainable energy projects are introduced into a community

[22–28]. Wind farm construction increases the loads on roads leading to and from the wind farm during turbine construction, but once the turbines have been constructed, there is nearly no increase in traffic [26, 27]. Conversely, when a fixed-point energy source that must be serviced by trucks is constructed, such as an ethanol or biomass plant, it results in additional traffic, and on many occasions, increased loads [22–25]. While it may be possible to mitigate these effects by the use of barge or rail [29], at some point, the road network will need to be used to move the turbine components, or the biomass or ethanol products.

In Indiana, ethanol plants, biomass power plants, and wind farms are typically built in rural areas. Most local road networks were not designed or constructed to accommodate the increased traffic and loads produced by such facilities. When sustainable energy developers decide to locate facilities within a given governmental entity, local officials need to have a sound understanding of the proposed facilities' probable effects on their local road network and some methods to quantify those effects. The local highway engineers and supervisors also need to be familiar with the resulting traffic and load problems associated with these facilities and be in a position to make decisions as to which pavement structure is needed to bear such heavy loads and traffic near the facilities.

Previous research in biomass and ethanol usage has pointed out the importance of designing access roads or considering the capacity of access roads to the plant [30, 31]. However, the existing literature, to the authors' knowledge, does not illustrate how loads can be calculated or access roads be designed for these facilities. Furthermore, research that has focused on wind farm technology suggests that local roads should handle the heavy construction loads from the wind mill parts [32–36]. On one hand, the focus has been on detailed design methodologies aiming to handle these loads, while on the other hand, guidelines to develop temporary access roads for wind farms were also presented [37].

As illustrated herein, the aforementioned design problems are solved using existing design guides, such as the American Association of State Highway and Transportation Officials (AASHTO) method. However, such methods may often be intricate, which would inevitably require consultation with expert designers. Even though third-party expert consultation is welcome, local authorities generally do not have the necessary funds for this process.

FIGURE 1: Ethanol plants in Indiana.

This paper aims to develop tools that can be used by local government agencies in quantifying the effects of proposed sustainable energy projects on their local road networks. The tools are designed and developed, bearing in mind that local agencies do not typically employ personnel with specific expertise in pavement analysis and design. These tools are therefore expected to assist local agency personnel in determining appropriate pavement sections and quantifying their costs. The paper is organized as follows. First, background information on renewable energy resources is given, along with biofuel transportation practices. Next, the method and data are presented, followed by the design development description of the proposed tool. Finally, the tool validation results are discussed.

The contribution of this paper lies in the development of local roadway infrastructure design guidelines associated with the construction and operation of sustainable energy source facilities, such as ethanol plants, biomass plants, and wind farm facilities. The proposed procedure is designed to be simple and is accompanied by hands-on tools to assist local government engineers in evaluating and in quantifying the probable effects of the construction and operation of a sustainable energy facility in their jurisdiction. Therefore, the procedure is anticipated to provide designs capable of withstanding heavy traffic loads, while, at the same time, it has the potential to quantify the effects that new sustainable energy source facilities may have on local road networks.

2.1.1 RENEWABLE ENERGY RESOURCES

In order to better comprehend the local effects of the construction and operation of sustainable energy projects, such as ethanol, biodiesel, biomass, and wind energy, some background information is briefly presented. Ethanol can be produced from a number of agricultural products, such as sugar and starch [38]. The ethanol production process yields several by-products, such as dried distillers grains with solubles (DDGS), which are a high-nutrient feed valued by the livestock industry [39]. Ethanol demand is difficult to capture, given its dual nature, i.e., being both an additive to and a substitute for gasoline. However, the market for ethanol significantly increased (over 500%) when flexible fuel vehicles (FFV) were made available to the public [21, 39, 40]. In Indiana, there are 11 ethanol plants, plus 2 under construction (see Figure 1).

Biomass is a plant matter grown to generate electricity or produce heat, with agricultural waste being the most common type of solid biomass that can be used as a source of energy [41]. Biomass currently provides about 10% of the world's primary energy supplies, most being used in developing countries in the form of fuel wood or charcoal for heating and cooking [42, 43]. In the USA, 85% of the wood production industry waste is used for power generation, with approximately 80 operating biomass power plants (and 40 operable but idle plants) located in 19 states across the country [44]. Demand for power derived from biomass is generally increasing, having surpassed hydropower as the largest domestic source of renewable energy [45]. In Indiana, there is currently one biomass plant near Milltown in Crawford County [46].

The power of the wind can be harnessed and converted to electricity by the use of tower-mounted wind turbines. Wind turbines can be used to produce electricity for a single home or building, or they can be connected to an electricity grid for more widespread electricity distribution. Wind energy is not only 'green' but also cost effective when compared to other sources of electricity in the USA. The growing wind power market has attracted many energy corporations to the field [47, 48]. In the USA, not all regions have wind speeds that are high enough to support wind energy production [49]. However, a recent study showed that building wind farms on only 3% of the area of the USA will produce enough electricity to meet all US energy demands [50]. In Indiana, there are currently 18 wind farms in operation, with over a 1,500 MW of wind electricity-generating capacity [51]. Indiana has the potential to produce 150,000 MW of electricity from wind farms [51].

2.1.2 TRANSPORTATION OF BIOFUELS

Biofuels can be transported by trucks, rail, or barge. Trucks are used when the material needs to be transported from one mode to another. Rail transportation is effective for long hauls, while barges are the least expensive transport method. Barges can carry large amounts to export terminals, and then ocean vessels are used to carry them to foreign markets. Transportation of biofuels using pipelines is limited in the USA due to the adverse impact of the former (mostly their chemical properties) on the pipeline integrity and safety. Also, pipelines are not largely available where biofuel plants are located. However, pipelines are a feasible option for the transportation of conventional fuel types. Air transport is not a viable option either, due to its high cost.

From a capacity standpoint, a truck can accommodate approximately 25,000 liters, a railcar approximately 95,000 liters, and a barge approximately 1,500,000 liters. On the other hand, it would not be economically effective to use rail or barges to transport biofuels to short distances, such as locations within less than 80-km distance from the facility [52]. In general, truck transportation is considered to be an efficient mode of transportation up to a distance of 500 km [52]. A railcar can transport the freight 2.5 times farther than a truck, for the same cost per liter, whereas barges can move freight across long distances and oversees (e.g., from the Midwest to the Gulf).

Rail is used to transport 41% of US corn exports and 14% of corn domestically [53]. In 2005, rail was the primary transportation mode for ethanol, shipping 60% of ethanol produced, or approximately 11 billion liters. In comparison, trucks shipped 30%, and barges 10%. Although trucks are used to ship most of the corn used by ethanol plants, some of the newer and larger plants use rail for inbound corn shipments [54].

Barges move approximately 5% to 10% of ethanol, in addition to the DDGS and fertilizers necessary for the production of corn. Barges also move 44% of all grain exports. In 2007, barges moved 55% of corn to ports, and 1% of corn to processors, feed lots, and dairies [54]. An issue with barge transportation is related to the occasional inadequacy of water depths that can lead to higher transportation costs. Seasonal effects on barge transportation may also decrease the barge's moving capacity (e.g., at a 2.75-m draft, if a barge has 1,500 tons of capacity, every 2.5 cm of reduced draft will result in 17 tons of reduced capacity) [53].

TABLE 1: Summary of local agency survey responses

County	Is the plant operational?	General response	Upgrades performed
Cass	Yes	No response	Unsure
Grant	No	No upgrades were performed on county roads. The nearest state highway was widened to accommodate the large-radius turning paths of long trucks. The plant does not have a county access road.	No
Henry	Yes	The plant is located right adjacent to a state highway; thus, there was no need for any upgrades. However, roads are deteriorating quickly, and there is no funding from the state or other sources.	No
Jasper	Yes	No provisions were needed because the plant is located adjacent to a state highway.	No
Jay	Yes	Upgrades were performed on county roads. The plant created a tax increment financing (TIF) district, and the new roads were paid for using the money from the bonds sold. Upgrades included widening and resurfacing of a section of a county road. The main problem is that truck drivers do not always use that route; thus, other roadways may deteriorate.	Yes
Kosciusko	Yes	The Highway supervisor expressed concern about the highways. Attempts were made to get funds to perform repairs. No legal agreement between the plant and the county was made	No
Lake	No	No response	No
LaPorte	No	No response	No
Madison	Yes	No response	Unsure
Montgomery	Yes	No response	No
Posey	Yes	There are two plants; one is adjacent to a state highway, the other is not. The latter required road upgrades. The upgrades were paid for through setting up a TIF district. Also, there were two low-volume roads that the plant wanted to build a bridge over. The county engineers were able to reach to an agreement with the commissioners to close these two roads, saving the expense of building an overpass. In return, the county received one million dollars which they used to repair and upgrade highways. The upgrades included mainly 5 to 7.5 cm of resurfacing on access roads.	Yes

TABLE 1: *Cont.*

County	Is the plant operational?	General response	Upgrades performed
Putnam	No	No response	No
Randolph	Yes	The plant built a private access road to a county road that was partly upgraded. The county is currently working on an agreement with the plant to upgrade the roads used by farmers.	Yes
Shelby	Yes	No response	Unsure
Wabash	Yes	The county established a TIF district in the area to be developed. The county performed road upgrades which included digging up the existing pavement, placing a 33- to 38-cm Portland cement-stabilized soil and HMA on top. The project cost was $1.2 million. The county was later reimbursed by the plant (as agreed before the start of the project by selling TIF district bonds). The county also received an economic development stimulus from the state of Indiana.	Yes
Wells	Yes	There were no upgrades performed. However, there were discussions at the time of construction that the plant had a budget set for upgrading the roadway. Due to technical difficulties on the county/city side, the roads were not upgraded. The plant did not spend any of the allocated budget. The county engineers tried to mitigate the damage by channelizing the truck traffic produced by the plant onto roadways that could accommodate the traffic. The county engineers provided this channelization through verbal coordination with truck companies and drivers. The highway supervisor stated that the truck companies were very cooperative.	No

In the Midwest, inbound corn being delivered to the processing facility is most typically delivered by trucks from corn farms within an 80-km radius. Standard gasoline tanker trucks (DOT MC 3066 Bulk Fuel Haulers) are typically used to ship ethanol outbound from the plants to the blending terminals. The total number of independently operated tank trucks is approximately 10,000, excluding the tanker truck fleets that are owned by petroleum companies [55].

2.2 METHODS

The objective of this paper is to develop a design methodology to assist local agencies in designing suitable pavements for sustainable energy projects served by local roads. In order to ensure a reliable pavement design, the first step is to collect accurate data with respect to the operation and traffic generation of sustainable energy projects. Development of the design tools is completed based on the following criteria: the tools should (a) be simple and easy to use; (b) require minimum input from the user and, at the same time, allow for more experienced users to input more detailed data; and (c) be able to produce several alternative pavement sections, when applicable. The output of the study involves worksheet-based pavement design procedures, one for ethanol and biomass plants, and a second for wind farms. These tools offer a user-friendly interface and several levels of input regardless of the expertise of the user.

The design development phase is based on various design guides and design elements that have been proven useful in the design of specialized pavements for sustainable energy projects. The design guides considered are the AASHTO Pavement Design Guide (for flexible and rigid pavements and for low-volume road design), the Asphalt Institute Pavement Design Guide, the Mechanistic-Empirical Pavement Design Guide (MEPDG), and the Portland Cement Association (PCA) [56–59]. Reviewing of these sources shows that rigid pavement design is not typically used in the design of local low-volume roads. Thus, the AASHTO Rigid Pavement Design Guide and the PCA Pavement Design were not utilized in the proposed overall design methodology. The MEPDG was found to be complex and was not geared toward low-volume roads and was therefore not used either. The AASHTO flexible pavement design was utilized due to its simplicity, versatility, and robustness. The AASHTO low-volume road design was also utilized in the ethanol and biomass worksheet. As for the wind farm worksheet, the Asphalt Institute's Manual Series No. 23 (MS-23), 'Thickness Design: Asphalt Pavements for Heavy Wheel Loads' was the only design guide that addressed the large, one-time loads expected during the construction of wind farm facilities [60].

As a final step, the proposed design methodology is tested to ensure that it produces realistic results. Ideally, the proposed design methodology would be validated by building a road conforming to the design methodology, then monitor it over several years, and determine whether it fails prematurely. Obviously, this falls out of the scope of the current study. Instead, the proposed methodology is validated by comparing it to in-service designs currently servicing sustainable energy projects. If the simplified proposed design provides an output that falls close to the outputs of the designs in place, the proposed design is considered adequate. This, of course, does not guarantee an optimum design; it suggests, though, that the developed designs are approximating actual design results.

2.1.1 INTERVIEWS WITH LOCAL OFFICIALS

As part of the data collection process, interviews were conducted with Indiana's local road agency representatives in counties where biomass plants, ethanol plants, or wind farm facilities are located. The interviews entailed a set of questions about the provided provisions in anticipation of the increased traffic and the current condition of the road network. To that end, 12 counties that have ethanol and biomass plants were interviewed. Of those 12, only 4 had performed any type of upgrade to their local roads in anticipation of increased traffic. Table 1 summarizes the representatives' responses.

Traffic associated with ethanol and biomass plants can be classified as follows: (a) incoming traffic handling raw materials and (b) outgoing traffic handling product distribution. Incoming traffic is mainly composed of trucks, while outgoing traffic is composed of rail and truck traffic, in most cases. For this reason, plants are typically located near major highways and rail sites. In Indiana, all plants are located within 4 km from the nearest state highway or interstate and within 1.1 km from the nearest rail freight facility. Of all the plants, 85% are located within 1.6 km of a state highway. Of all operating ethanol and biomass plants in Indiana, 30% are located adjacent to a main highway, whereas 23% of all operating ethanol and biomass plants in Indiana have rail tracks leading into their facility. On average, the plants in Indiana are 0.87 km away from a state highway and 0.5 km from a railroad.

County officials and plant managers expected that all truck traffic would use the nearby state highways or interstates. Thus, no significant upgrades were performed on local roads. In many cases, the expectation that trucks would utilize the state highways or interstates was not validated. Truck drivers use the shortest route unless otherwise instructed, which may or may not be a state highway or interstate. Also, farmers delivering raw materials to the plant came from all directions. This entailed utilizing county roads.

Local county highway representatives were interviewed in several counties in Indiana, in which wind farms are located. In both cases, the wind farm developers signed a road use agreement with the county specifying that the developers are responsible for the road condition. The developers agreed to return the roadways used in the wind farm construction process to their original condition and further performed significant upgrades to the local roads. However, detailed information was only available from White County, which was used in the validation process.

Note that road use agreements typically include warranty clauses, which provide an assurance to the owner that the product/service will serve its useful life without failure, and if it does not, the contractor will repair or replace the product (for specifics on roadway preservation through public-private partnerships, see [61–69]). In the case of White County, a 2-year warranty was defined. Benton County defined a 1-year warranty on roads and a 5-year warranty on drainage.

2.2 DATA

2.2.1 BIOMASS AND ETHANOL PLANTS

The amount of traffic associated with an ethanol plant is directly related to the plant's capacity, most often measured in millions of liters per year (MLY). Because the plants are normally located to take advantage of locally produced raw materials, in this case corn, nearly all of the incoming raw material is delivered to the plant by tractor-trailers. The outgoing products are ethanol and DDGS. In Indiana, nearly all of the ethanol leaves the plant by train. The DDGS may be transported by train or truck, depending on local livestock markets. Plant capacities, amount of raw materials consumed, and plant production rates for each of the ethanol plants in Indiana are summarized in Table 2.

TABLE 2: Indiana ethanol plant data

Plant	County	Annual liters of ethanol produced (millions)	Annual bushels of corn used (millions[a])	Corn used per liter of ethanol produced (bushels)	Annual tons of DDGS produced (thousands)	Annual tons of DDGS produced per liters of ethanol produced (millions)
Anderson Ethanol	Cass	416	39	0.0937	354	0.850
Cardinal Ethanol	Randolph	379	37	0.0977	321	0.848
Central Indiana Ethanol	Grant	151	15	0.0991	145	0.958
Indiana Bio-Energy	Wells	416	37	0.0889	321	0.771
Iroquis Bio-Energy Company	Jasper	151	15	0.0991	129	0.852
New Energy Corp.	St. Joseph	379	37	0.0977	328	0.866
POET	Jay	246	24	0.0975	193	0.784
POET	Madison	227	22	0.0969	193	0.850
POET	Wabash	246	24	0.0975	209	0.849
Valero Energy (formerly Vera-Sun)	Montgomery	379	37	0.0977	350	0.925
Altra (not operating)	Putnam	227	22	0.0969	192	0.845
Abengoa Bio-energy	Posey	333	32	0.0961	282	0.847
Total		3,551	341	1.16	3,017	10.25
Average		296	28	0.10	251	0.85

[a]*One bushel of corn weighs about 25 kg.*

The amount of raw material consumed by a biomass plant is governed by the plant's capacity, the amount of electricity it can produce, and the plant's efficiency. Capacity is measured in megawatt electrical (MWe), while efficiency by the heat production rate is measured in watts per kilowatt-hour (W/KWh). Each material, when burned, produces a specific amount of heat energy measured in watts per kilogram (W/kg). Herein, a constant value of 2,746 W/kg for all agricultural byproducts is adopted from Wiltsee [70]. The average heat rate of 140 biomass plants listed in the National Electric Energy System Database [71] was also used, which was calculated to be 4,462 W/KWh.

Unlike ethanol plants, biomass plants do not produce loaded, outgoing traffic. Raw materials are shipped to the plant and burned to generate electricity. The type of input materials varies and can be divided into four main types: woody plants, herbaceous plants/grasses, aquatic plants, and manures [72]. According to the local Indiana farmers, woody and herbaceous plants are the most commonly used raw materials in Indiana biomass plants, with the most typical being corn stover, wood chips, sawdust, and baled straw [72–74]. Each material has a different density, as shown in Table 3. The less dense the material, the more space per kilogram it occupies; thus, more trucks are needed to transport less dense materials. This was taken into consideration when calculating loads associated with biomass plant operation.

TABLE 3: Biomass raw material densities

Material	Density (kg/m^3)
Corn stover	128.15
Wood chip	200.23
Sawdust	120.14
Baled straw	150.57

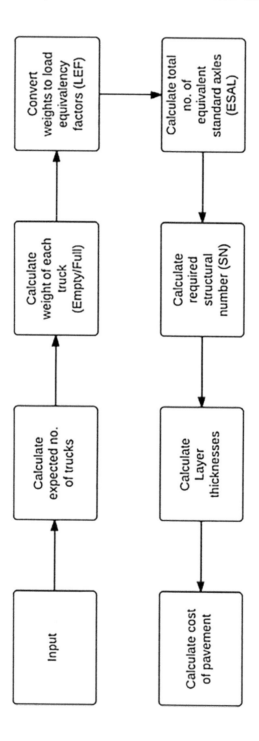

FIGURE 2: Conceptual illustration of the pavement design procedures.

TABLE 4: Truck information for various wind turbine components

Component	Weight (kg)	Longest dimension (m)	Minimum number of truck axles needed to carry component	Weight per axle (kg/axle)	Weight per tire (kg/tire)
Base section	41,958	14.66	4	10,490	2,622
Lower-middle section	41,241	19.81	6	6,874	3,789
Upper-middle section	28,111	19.90	6	4,685	2,583
Top section	28,876	22.59	6	4,813	2,653
Hub	17,010	3.84	3	5,670	3,125
Blades	6,486	33.99	6	1,081	596
Rotor	32,024	70.47	6	5,337	2,942
Nacelle	57,153	8.81	3	19,051	10,500

2.2.2 WIND FARMS

The increased truck traffic associated with wind farm facilities is mostly limited to construction traffic, which can be divided into transportation of construction materials (concrete, aggregates, and steel reinforcing), transportation of construction equipment (cranes), and transportation of wind turbine components (nacelle, rotor, blades, and tower sections). The construction materials represent the heaviest loads per truck axle. The turbine components can be heavy, but additional axles are added to the truck trailer as needed so as to comply with axle weight limits. In most cases, the length of the turbine components is the most critical concern. Wind turbine components, such as blades and tower sections, are extremely long and require long trucks to haul them. Blades are typically 45 m in length and weigh 11,340 kg [26]. While the weight is distributed over a large number of axles, the challenge is making sure that trucks have sufficient turning radii when using local roads. Table 4 summarizes the weight and truck axles needed for various wind turbine components [71].

The erection of wind turbines includes two major activities: off-loading and stacking out. Off-loading normally requires a 200-ton crawler or hydraulic crane. Stack-out requires a 400-ton crane [75]. Both cranes are transported in pieces and assembled on site. Table 5 summarizes the weight of each component and the number of truck axles required to carry it. Each crane is assembled in 20 to 25 truck trips, which are performed at least twice (assembling and disassembling) in the project lifetime, regardless of the number of wind turbines being built [26, 76, 77].

TABLE 5: Crane components

Equipment	200-ton crane			400-ton crane		
	Weight (kg)	Number of axles	Weight per tire (kg/tire)	Weight (kg)	Number of axles	Weight per tire (kg/tire)
Basic crane	39,689	4	2,480.58	39,612	4	2,475.76
Car-body and adapter	N/A	N/A	N/A	28,161	3	2,346.77
Crawler assembly	19,622	4	1,226.40	32,665	4	2,041.59
Counterweight tray	9,548	3	795.68	19,958	3	1,663.17
Upper-center counterweight	10,659	3	888.28	8,165	3	680.39
Upper-side counterweight	7,938	3	661.49	6,804	3	566.99
Lower car-body counterweight	9,979	3	831.59	13,608	3	1,133.98
Upper car-body counterweight	8,165	3	680.39	N/A	N/A	N/A
9 m boom butt	4,910	3	409.18	21,609	3	1,800.76
12 m boom top	2,544	3	212.02	5,595	3	466.25
3 m boom insert	971	380.89	N/A	N/A	N/A	
6 m boom insert	1,397	3	116.42	2,563	3	213.57

As mentioned earlier, the heaviest load associated with wind farm construction is the construction materials. Wind tower foundations require 282 to 480 yd³ of concrete and 20 to 38 tons of steel reinforcement [75]. Truck traffic is also generated by the need to transport aggregates to the site. Table 6 presents the number of trucks needed to construct the foundation of a single turbine and the weights of each construction material used

[23]. Finally, data on local pavement construction materials were collected from local suppliers and used in the proposed design process. One of the design outputs is the cost of the recommended pavement. The specific gravity and cost data collected and utilized in the analyses are illustrated in Table 7.

TABLE 6: Wind-tower foundation construction materials

Construction material	No. of trucks required	Truck loads (kg)	Load per tire (kg)
Aggregate	10	22,680	5,670
Concrete	20 to 40	22,680	5,670
Steel	1	18,144	4,536

TABLE 7: Pavement construction materials, specific gravities, and costs

Material	Specific gravity	Density (kg/m³)	Tons/lane-km/cm	Price/ton/lane	Price/lane-km/cm
Hot mix asphalt	2.65	2,643	106.66	$90	$9,599
Compacted dense aggregate	2.75	2,739	110.82	$13	$1,441
Coarse aggregate	2.45	2,435	98.83	$9	$889
Excess excavation					$245

2.3 RESULTS AND DISCUSSION

2.3.1 DESIGN DEVELOPMENT

Two MS Excel-based pavement design procedures (Excel worksheets) were developed, one for ethanol and biomass plants, and a second for wind farms. Both worksheets follow the logic outlined in Figure 2.

TABLE 8: Primary and secondary inputs needed for the ethanol and biomass procedure

Primary inputs	Description	Secondary inputs	Description
Plant capacity	This value represents the maximum amount of biofuel that could be produced by the plant considered. This value should be reported in million liters per year for ethanol plants and megawatts electrical (MWe) for wind farms.	Yearly growth factor	If the plant is expected to increase its capacity in the future, the designer may add a reasonable growth factor. However, it is uncommon for a plant to be built and have its capacity increased later on in its service life. The default value is 0%.
Capacity factor for biomass plants	Biomass plants produce power but are not exclusively consistent in doing so. The capacity factor is a percentage that represents the average power output of a plant. It ranges from 15% to 100%. When this value is unknown, an average value of 67% is assumed.	Ethanol plant products and raw materials	Amount of corn hauled to the plant by trucks, as opposed to be transported by another means, or grown within the plants' grounds, thus not using any local roads. Therefore, the default value assigned is 100%.
			Ethanol hauled from the plant by trucks, as opposed to stored or sold locally. Most ethanol plants are located close to rail freight tracks. For this reason, most (if not all) of the ethanol production will be shipped by rail. The assigned default value is conservatively set to 20%.
			DDGS hauled from the plant by trucks. This has a default value of 20%.
Design period	The amount of time the road is expected to remain in service without major rehabilitation. This value is typically 20 years; however, for county roads, it can be lower.	Biomass fuel type	Biomass fuel can be produced using various components and ratios of these components. Different component raw materials have different weights. It is important to determine the different percentages of each raw material to avoid underestimating the weight of incoming trucks. The four typical components are corn stover, wood chips, saw dust, and baled straw. Each is set at a default value of 25%.

TABLE 8: *Cont.*

Primary inputs	Description	Secondary inputs	Description
California bearing ratio (CBR)	This value reflects the strength of the underlying soil. To get the actual field CBR of the soil, soil bores need to be drilled in the construction location. However, the value can be closely estimated by knowing the type of soil in the area. Highway supervisors can resort to previous experience or soil maps to determine the soil type in the area.	Reliability (R%)	The designer should choose the level of reliability of the design. For local county roads, the reliability is typically low. This value ranges from 50% to 99%; 75% is the default value.
		Terminal serviceability index (Pt)	This is the value that reflects the condition of the pavement at the end of its service life. This value ranges from 3 (for major highways) to 1.5 (minimum). The default value is 2, as recommended for county roads by the AASHTO design guide [49].
		Overall standard deviation	This number reflects the variability within the pavements' materials. It typically ranges from 0.3 to 0.5. A value of 0.5 is recommended by the AASHTO design guide [49] and is set as the default value.

2.3.2 ETHANOL AND BIOMASS DESIGN SPECIFICATIONS

The ethanol and biomass design procedure is based on the standard AASHTO Pavement Design Guide [56]. As discussed previously, ethanol and biomass plants produce significant traffic, which is mainly composed of trucks. The traffic produced is typically above one million equivalent single axle loads (ESAL) over a 20-year period, which merits a design governed by pavement fatigue standards.

There are primary and secondary design inputs, which are summarized in Table 8. Primary inputs are project-specific, which vary significantly from one project to another. Secondary inputs are approximately constant across projects, either due to the nature of these factors, or because they tend to be standardized. These inputs are set at default, but can be customized by the user if desired.

In order to simplify the design process, a number of assumptions are made. It is important to note that these assumptions could be easily changed by the user if so desired. The following are the assumptions considered in this design procedure:

- The capacity of a typical truck is assumed to be 90.6 m³.
- An average number of bushels of corn (0.091 bushels/liter of ethanol) is used to calculate the amount of corn needed to supply the ethanol plant:
- *Unit analysis (1 bushel = 25 kg)*
- *For one million liters of ethanol, 91,000 bushels are needed (0.091 × 1,000,000); using a 25-ton (25,000 kg) truck, the number of trucks needed per day in a year is equal to 91,000 × 25 / (25,000 × 365) = 0.25.*
- Using unit analysis, it is found that for each one million liters per year of ethanol to be produced, 0.25 trucks per day are needed to supply the plant. The number of ethanol tanker trucks is calculated, assuming that tankers haul on average 30,000 liters. Using an ethanol density of 0.789 kg/liters, the weight per truck is calculated as 23,000 kg [55].
- One million liters of ethanol has a byproduct of 845 tons of DDGS and a truck can haul approximately 25 tons of distiller's grain [55], leading to a factor of 0.0926 trucks per day:
- *Unit analysis*
- *For one million liters of ethanol, 845 tons of DDGs is needed; using a 25-ton (25,000 kg) truck, the number of trucks needed per day in a year is equal to: 845 / (25 × 365) = 0.0926.*
- The ESAL of the trucks are calculated using the fourth power law load equivalency factor (LEF). This law uses the weight over a constant based on axle type raised to the fourth power [78]:

$$\text{LEF} = (\text{weight of axle/weight of constant})^4 \tag{1}$$

$$\text{ESAL} = \text{LEF} \times \text{number of vehicles with that axle weight} \tag{2}$$

For a single axle, the constant used is 8,160 kg. For a tandem axle, the constant used is 15,060 kg. where W_{18} = ESAL (reflects road traffic), Z_R = standard normal deviation (reflects the reliability of design), S 0 = standard deviation (reflects variability of pavement material), ΔPSI = reflects the difference between pavement condition right after construction and the end of its service life, and M_R = resilient modulus (PSI) (reflects subgrade strength).

- For the biomass facility, the weight of product produced is calculated using typical heat and production rates: 4,103 W/kWh heat rate and 2,746 W/kg fuel production rate.
- For biomass facilities, the densities of the raw materials are used to calculate the weight and number of trucks going into the plant. The densities of these raw materials are listed in Table 3.
- The AASHTO design guide's [56] ESAL equation is used to find the pavement's structural number:

$$\log W_{18} = Z_R + S_0 + 9.36(\log(SN + 1) - 0.2) + \frac{\log \dfrac{\Delta PSI}{4.2 - 1.5}}{0.4 + \dfrac{1,094}{(SN + 1)^{5.19}}}$$

$$+ 2.32(\log(M_R) - 8.07)$$

(3)

The design normally consists of three layers. Layer coefficients, a, are assumed to take the following values: a_1 = 0.4 (in the range of 0.2 to 0.4; 0.4 is typically used), a_2 = 0.14 (recommended by AASHTO for granular base layers), and a_3 = 0.11 (recommended by AASHTO for granular subbase layers). Drainage coefficients, m, are assumed to take the following values: m_2 = 0.8 and m_3 = 0.8 (these values reflect fair drainage with 75% reliability). Initial pavement serviceability (Po) is assumed to be 4.2 (this value is the AASHTO design guide [56] recommended value for flexible pavements). The resilient modulus introduced in the design's internal calculations is a function of the California bearing ratio or CBR (a penetration test for the evaluation of the mechanical strength of road subgrades and base courses, with higher values representing harder surfaces) and is calculated using the following equation:

$$M_R = 2,555 \, (CBR^{0.64}) \tag{4}$$

- The assumed specific gravities and costs of materials are listed in Table 7. Note that these costs are highly recommended to be updated regularly to reflect actual costs.

Table 9 provides a summary of all input parameters, default values, and assumptions.

TABLE 9: Ethanol and biomass worksheet input parameters

Parameter	Default value	Range	Comments
Plant capacity	User input	N/A	Input in MLY or MWe
Design period (years)	20	1 to 40	A service life of 20 years should be adequate for low-volume pavements.
Yearly growth factor (%)	0	0 to 100	This is capacity growth associated with the plant alone. It is not expected that a plant at full capacity will increase its capacity with time; thus, this value is set at 0%.
Ethanol plants only			
Percentage of corn used that is trucked to the plant (%)	100	0 to 100	The model assumes all corn is supplied to the plant by trucks.
Ethanol leaving the plant by truck (%)	20	0 to 100	The model assumes 20% ethanol is hauled from the plant in trucks.
Dried distillers grain leaving the plant by truck (%)	20	0 to 100	The model assumes most of the dried distiller grains leave the plant by rail.
Biomass plants only			
Capacity factor (%)	66.6	19 to 100	The average capacity factor of biomass plants across the USA is 66.6%.
Fuel type for biomass			
Corn stover (%)	25	0 to 100	Each plant uses different raw materials with different densities. This, in turn, affects the number of trucks supplying the materials. Currently, there is no dominant material in Indiana, thus the equal division.

TABLE 9: *Cont.*

Parameter	Default value	Range	Comments
Wood chips (%)	25	0 to 100	
Sawdust (%)	25	0 to 100	
Baled straw (%)	25	0 to 100	
Structural parameters			
California bear ratio	User input	N/A	CBR is then converted to resilient modulus (M R) using Equation 4; a value of 3 or less can be used for conservative results.
Standard normal deviate (based on percent reliability), Z_R	75	50 to 99	The designer should choose the level of reliability of the design. For local county roads, the reliability is typically lower than high-volume roads.
Terminal serviceability, P_t	2	3.0 to 1.5	This is the value that reflects the condition of the pavement at the end of its service life. The AASHTO design guide recommends a value of 2 for county roads.
Standard deviation, S_o	0.5	0.3 to 0.5	This number reflects the variability within the pavements' materials. A value of 0.5 is recommended by AASHTO design guide.

For ethanol plants, the number of trucks carrying products and raw material (or empty) is determined by multiplying the amount of ethanol produced per year (in MLY) by the number of trucks needed to ship one million liters of ethanol in a day (i.e., 0.25 for ethanol, and 0.0926 for the added value product DDG) or dividing by the average liters of ethanol carried by each truck daily (30,000 liters/truck). The weight of each truck is calculated using the weight of the truck empty plus the weight of the cargo (raw materials or products carried). The steering axle has a fixed load. Driver and trailer axles share the weight of the load equally. The loads are converted to load equivalency factors (LEF) using Equation 1.

In biomass plants, raw materials are burned to produce electricity. Each kilogram of raw material can produce a certain amount of heat energy. Typically, 1 kg of biomass can produce 2,746 W of heat energy; this is labeled as the production rate. The heat is then used to convert liquid water into steam, which in turn, rotates a steam turbine to produce electricity. The amount of heat needed to produce 1 KWh of electricity is the heat rate of the process. The heat rate is typically 4,103 W/KWh. The amount of material needed per day can be obtained by multiplying the heat rate by the number of KWh produced in a day and then dividing by the production rate.

The next step is to calculate the number of trucks needed to carry the raw material. First, the weight of each raw material is calculated; next, it is divided by its density and converted into volume. The total number of trucks is then calculated by dividing the total volume of that material by the capacity of each truck, which is typically 90.6 m3. The weight of each truck can be obtained by multiplying the total truck capacity (90.6 m3) by the density of the raw material used. The raw material used by the worksheet can be corn stover, woodchips, sawdust, and baled straw, or any combination of these. The loads are converted to LEF using Equation 1.

In both ethanol and biomass worksheets, the total number of ESAL is calculated by multiplying the LEF by the number of trucks in the design period that have that axle, and then by summing up all the ESAL.

The structural number (SN) for all layers is calculated using Equation 3. The thickness of each layer is calculated using the attained structural number, and the layer and drainage coefficients. The AASHTO [56] structural number equation is

$$SN = a_1 D_1 + a_2 D_2 m_2 + a_3 D_3 M_3 \qquad (5)$$

For full-depth asphalt, the depth of the layer is obtained by dividing SN by a 1. The worksheet calculates various layer thickness combinations. The first layer is initially set to the minimum recommended by AASHTO [56] for the number of ESALs attained. The base and sub-base are calculated by satisfying two simultaneous equations. The first is the structural number equation, and the second is the ratio of base to sub-base thickness set by the user. The tool computes various combinations of the three layer thicknesses: D_1, D_2, and D_3. The user can chose any combination or change the values to produce a unique design. For full-depth asphalt, the depth of the layer is obtained by dividing SN by a 1.

The worksheet automatically estimates the costs for the thickness combinations using the assumed cost values listed earlier. However, the users have the option of specifying their own costs.

2.3.3 WIND FARM DESIGN SPECIFICATIONS

Wind turbines have very large and heavy components. These components, when transferred to the wind farm location, can accelerate the deterioration of the road assets and their components. Similar to the ethanol and biomass procedure, there are two levels of input, primary and secondary. These are presented in Table 10.

TABLE 10: Primary and secondary inputs needed for the wind farm procedure

Primary inputs	Description	Secondary inputs	Description
Number of wind turbines	To transfer each wind turbine, a certain number of trucks are required, which typically use the same transfer routes. In the case that several turbines are transferred by different routes, each route should be designed for the number of turbines that will be moved across.	Tire contact area	Different tires have different contact areas. This value could be obtained from the manufacturer. The load and internal pressure are factors that affect and are affected by these values. If this value is unknown, 1,935 cm^2 is recommended as default.
California bearing ratio (CBR)	This value reflects the strength of the underlying soil. To get the exact value, soil bores are needed to be drilled in the construction location. However, the value can be closely estimated by knowing the type of soil in the area. Highway supervisors can resort to previous experience or soil maps to determine the soil type in the proximity.		

To simplify the design process, a number of assumptions are made. It is important to note that all these assumptions could be changed manually in the design guide spreadsheets. The following are the assumptions considered in the wind farm design procedure:

- One-time heavy wheel loads during construction: Use Asphalt Institute's manual 'Thickness Design: Asphalt Pavements For Heavy Wheel Loads'. The loads necessary for a single wind turbine to be built (components and construction) are shown in Tables 4 to 6.
- Turbine loads based on a GE 1.5 s (1.5-MW design): Loads are assumed to be similar for all turbines near this size.
- The resilient modulus introduced in the design's internal calculations is calculated using Equation 4.
- The specific gravities and cost of materials assumed are listed in Tables 6 and 7.

Table 11 lists all input parameters and their default values, where applicable.

TABLE 11: Wind farm worksheet input parameters

Parameter	Default value	Comments
Number of turbines	User input	The number of wind turbines to be installed.
Soil CBR	User input	CBR is converted to resilient modulus (M R) using Equation 4.
Tire contact area (cm²/tire)	1,935	This value could be obtained from the manufacturer. One thousand nine hundred square centimeters was used for dual-tire configurations [72, 73].
Maximum load per tire (kg)	4,536	Only construction loads are considered.

Traffic counts are calculated by multiplying the heaviest truck used for design, by the number of trucks per wind turbine (assumed 91), and by the number of wind turbines (inserted by the users). To estimate the layer structural numbers, two values are needed to be calculated: the tire coefficient, a, and the tire pressure. The tire contact area coefficient, a, is calculated as follows:

$$a = \sqrt{(tire\ contact\ area/\pi)}$$

(6)

whereas the tire pressure, p, is calculated as

p = maximum load/tire contact area   (7)

The structural number is calculated as

$$SN = 0.3a\left(0.772\ln(p) - 2.535 + \frac{15 - M_R}{10.5}(0.049\ln(p) + 0.116)\right)$$

(8)

To estimate costs of various alternatives, the cost of the calculated thicknesses is computed using the assumed cost values listed in Table 7 (the worksheet does this computation automatically); however, the users have the option of including their own cost values.

2.3.4 VALIDATION

The ethanol and biomass worksheets are based on the AASHTO pavement design guide [56], inheriting its strong points and its limitations. For the purposes of developing a user friendly design procedure, the AASHTO design method is preferred over the more recent mechanistic-empirical design method, as the former is well known for its empirical approach. Moreover, it does not need calibration or validation, because it is linked to the validated AASHTO design guide. For more conservative results, the reliability factor (ranges from 50% to 99.9%) can be increased in the worksheet.

Finding structural data associated with local roads is a tedious task. However, Jay, Posey, and Wabash counties did collect the pavement layer

thickness values after they were upgraded for the construction and operation of the biomass and ethanol plants. For the ethanol plant in Jay County,
provisions were made to accommodate the new increased truck traffic,
and the road was resurfaced; however, the assumptions made related to
the preferred truck-driver routes fall short, causing excessive deterioration
to the adjacent roads. For the ethanol plant in Posey County, to handle the
excess traffic, the county resurfaced all access roads. Similarly, to accommodate the anticipated increase in traffic associated with the ethanol plant
in Wabash County, a 1.2-km road section was reconstructed. With these
exceptions, the interviews with various county officials showed that there
were generally no provisions made for ethanol and biomass plants due to
their close proximity to state roads.

TABLE 12: Ethanol and biomass worksheet results

Design	Alternative 1	Alternative 2	Alternative 3	Alternative 4
Jay County, POET Plant, capacity 65 MLY				
Surface layer (cm)	7.62	8.89	10.16	10.16
Base layer (cm)	26.67	15.24	30.48	19.05
Sub-base layer (cm)	34.29	40.64	15.24	30.48
Pavement cost ($)	246,744	247,468	262,624	259,417
Posey County, Abengoa Bioenergy Plant, capacity 88 MLY				
Surface layer (cm)	7.62	8.89	11.43	10.16
Base layer (cm)	26.67	15.24	30.48	21.59
Sub-base layer (cm)	35.56	44.45	15.24	30.48
Pavement cost ($)	249,060	254,417	282,769	266,306
Wabash County, POET Plant, capacity 65 MLY				
Surface layer (cm)	7.62	8.89	12.7	10.16
Base layer (cm)	29.21	15.24	30.48	25.4
Sub-base layer (cm)	38.1	49.53	15.24	30.48
Pavement cost ($)	260,082	263,182	302,414	276,160

The second step in the validation process is to produce designs for
these three counties using the developed worksheets. Table 12 presents the
recommended pavement sections for each plant based on the ethanol and

biomass worksheet, along with an estimated cost for each alternative. A CBR value of 3, representing tilled farmland, was used for Jay and Posey counties, and a value of 2 (representing softer surfaces) was used for the Wabash County. The county highway engineers mentioned that the soil is weak in that particular area. A design period of 20 years was assumed in all three cases.

The final step in the validation process is to compare the pavements designed by the county engineers and private contractors, with the pavement designs proposed by the developed tool. Table 13 lists the capacities of various plants, and compares as-built thickness to those proposed by the worksheets. The structural numbers obtained by the worksheet are higher than the actual numbers from both cases. Wabash County pavements were designed by an engineering consulting firm. The other counties used developer and local suggestions. The results of the worksheet are closest to the engineering firm's recommendations. With all inputs (except the plant capacity and the design period) being set at default values, this suggests that a proper design may be consistent with the worksheet's default output. If more information was available, such as the soil CBR or the design period, the results would likely be more accurate.

Turning to the wind farm spreadsheet validation, it is noteworthy that in 2008, several lease agreements were signed with local farmers and land owners in White County, with the intention of building electricity-generating wind towers. A private consultant was hired to insure that the adjacent roads would be capable of handling the transportation of the wind tower components. The engineering firms performed an extensive field evaluation which included soil borings from all adjacent pavement sections, lab testing and soil classification of the collected soil samples, and performing non-destructive pavement testing using a falling weight deflectometer (FWD). Note that pavement surface deflection is typically used to evaluate the flexible pavement structure and the rigid pavement load transfer and is measured as the pavement surface's vertical deflected distance as a result of an applied static or dynamic load [79–82]. The FWD is the most common type of equipment to measure the surface deflection in Indiana, and the units used are thousandths of centimeters from a FWD center-of-load deflection, corrected to a 11,340-kg load applied on a 30-cm-diameter plate, adjusted for temperature (18°C) [83, 84].

The engineering firms along with the White County highway department concluded that pavement upgrades were required. The engineering firm assumed the construction of 127 wind towers (phase I of the project) which included 5,000 concrete trucks, 8,000 gravel haul trucks, 1,150 semi-trucks for turbine component delivery, and numerous passes by medium and heavy cranes. This resulted in a total of 14,150 vehicles (plus crane passes). This value is more conservative than the 11,557 vehicles assumed by the wind farm worksheet. Even though the worksheet underestimates the number of trucks, it is important to note that the design methodology is based on the maximum truck weight for any truck category that constitutes more than 10% of the truck traffic. The consulting firm and the worksheets both used 4,536 kg per tire as their maximum weight.

The engineering consulting firm developed a pavement design that included the pavement layer thicknesses to carry the wind turbine components and construction materials. The SN was attained by the consultant using the AASHTO design guide for low-volume aggregate surfaced roads [56] and considered allowable rutting. Table 14 lists the consultant and worksheet results.

TABLE 14: Wind farm worksheet and county consultant inputs and results

Input criteria	Consultant design	Worksheet design
Number of turbines	127	127
Truck traffic assumed	14,150	11,557
California bearing ratio	3	3
Soil resilient modulus (Mpa)	41.37	35.58
Allowable rutting (cm)	5 to 7.5	N/A
Allowable loss of service	3	N/A
Aggregate base modulus (Mpa)	206.4	N/A
Percent heavy trucks (%)	80	100
Maximum load per tire (kg)	4,536	4,536
Recommended structural number	1.3	1.9
Recommended pavement layer thickness	-	-
Hot mix asphalt surface (cm)	0	5.08
Stabilized aggregate base (cm)	30.5	30.5

As expected, the worksheet provides a higher SN than the one proposed by the consultant. This could be due to the fact that the consultant has more accurate information on this specific project due to the tests that were performed, or it could be due to the selection of design procedure. It can be argued that a low-volume road design is also applicable because the number of ESAL is expected to be small. However, the developed worksheet uses the heavy wheel design, which more closely matches the given traffic scenario (the heavy construction loads).

2.4 CONCLUSIONS

Data associated with sustainable energy facility traffic (such as number, type, and weight of trucks with or without cargo) were collected, to develop Excel-based tools (worksheets) and assist local agencies in the design of pavements in the proximity of ethanol plants, biomass plants, and wind farms. The worksheets provide a user-friendly environment for engineers with any level of expertise to produce a pavement design for the aforementioned facilities in an easy and timely fashion. Experienced designers have the option to change the default values of the worksheets in order to produce more cost-effective designs. Otherwise, the worksheets' default values can be maintained and still provide a conservative design.

From the comparison of the worksheet-generated designs and those practically implemented, it was found that the worksheet-proposed pavements were slightly thicker than the actual implemented designs, and thus less likely for the pavement to fail. This could reflect a need for collection of additional data points, or for further calibration of the tool through additional validation tasks. To that end, the as-built pavement sections will be revisited after 1- to 5-year intervals to assess their condition and further validate the worksheet tools.

The developed worksheets can serve as a hands-on tool to assist local government engineers in evaluating and quantifying the probable effects of the construction and operation of a sustainable energy facility in their jurisdiction. Further recommendations to assist in achieving this goal involve inclusion of biodiesel plants, further validation of the worksheets using measures of pavement distress (rutting or cracking), and comparison of the design outputs with actual data from constructed roads.

REFERENCES

1. Bischoff A: Insights to the internal sphere of influence of peasant family farms in using biogas plants as part of sustainable development in rural areas of Germany. Energ Sustain Soc 2012, 2:9. 10.1186/2192-0567-2-9
2. Dampier JE, Shahi C, Lemelin R, Luckai N: From coal to wood thermoelectric energy production: a review and discussion of potential socio-economic impacts with implications for Northwestern Ontario, Canada. Energ Sustain Soc 2013, 3:11. 10.1186/2192-0567-3-11
3. Galich A, Marz L: Alternative energy technologies as a cultural endeavor: a case study of hydrogen and fuel cell development in Germany. Energ Sustain Soc 2012, 2:2. 10.1186/2192-0567-2-2
4. Green JS, Geisken M: Socioeconomic impacts of wind farm development: a case study of Weatherford, Oklahoma. Energ Sustain Soc 2013, 3:2. 10.1186/2192-0567-3-2
5. Grunwald A, Rösch C: Sustainability assessment of energy technologies: towards an integrative framework. Energ Sustain Soc 2011, 1:3. 10.1186/2192-0567-1-3
6. Hagen Z: A basic design for a multicriteria approach to efficient bioenergy production at regional level. Energ Sustain Soc 2012, 2:16. 10.1186/2192-0567-2-16
7. Halder P, Weckroth T, Mei Q, Pelkonen P: Nonindustrial private forest owners' opinions to and awareness of energy wood market and forest-based bioenergy certification - results of a case study from Finnish Karelia. Energ Sustain Soc 2012, 2:19. 10.1186/2192-0567-2-19
8. Hassan MK, Halder P, Pelkonen P, Pappinen A: Rural households' preferences and attitudes towards biomass fuels - results from a comprehensive field survey in Bangladesh. Energ Sustain Soc 2013, 3:24. 10.1186/2192-0567-3-24
9. Klagge B, Brocke T: Decentralized electricity generation from renewable sources as a chance for local economic development: a qualitative study of two pioneer regions in Germany. Energ Sustain Soc 2012, 2:5. 10.1186/2192-0567-2-5
10. Mohr A, Bausch L: Social sustainability in certification schemes for biofuel production: an explorative analysis against the background of land use constraints in Brazil. Energ Sustain Soc 2013, 3:6. 10.1186/2192-0567-3-6
11. Niemetz N, Kettl K-H: Ecological and economic evaluation of biogas from intercrops. Energ Sustain Soc 2012, 2:18. 10.1186/2192-0567-2-18
12. Nishimura K: Grassroots action for renewable energy: how did Ontario succeed in the implementation of a feed-in tariff system? Energ Sustain Soc 2012, 2:6. 10.1186/2192-0567-2-6
13. Oyedepo SO: Energy and sustainable development in Nigeria: the way forward. Energ Sustain Soc 2012, 2:15. 10.1186/2192-0567-2-15

14. Palmas C, Abis E, von Haaren C, Lovett A: Renewables in residential development: an integrated GIS-based multicriteria approach for decentralized micro-renewable energy production in new settlement development: a case study of the eastern metropolitan area of Cagliari, Sardinia, Italy. Energ Sustain Soc 2012, 2:10. 10.1186/2192-0567-2-10

15. Scheer D, Konrad W, Scheel O: Public evaluation of electricity technologies and future low-carbon portfolios in Germany and the USA. Energ Sustain Soc 2013, 3:8. 10.1186/2192-0567-3-8

16. Stoeglehner G, Niemetz N, Kettl K-H: Spatial dimensions of sustainable energy systems: new visions for integrated spatial and energy planning. Energ Sustain Soc 2011, 1:2. 10.1186/2192-0567-1-2

17. Taheripour F, Hertel TW, Liu J: The role of irrigation in determining the global land use impacts of biofuels. Energ Sustain Soc 2013, 3:4. 10.1186/2192-0567-3-4

18. Taheripour F, Zhuang Q, Tyner WE, Lu X: Biofuels, cropland expansion, and the extensive margin. Energ Sustain Soc 2012, 2:25. 10.1186/2192-0567-2-25

19. Tunç M, Pak R: Impact of the clean development mechanism on wind energy investments in Turkey. Energ Sustain Soc 2012, 2:20. 10.1186/2192-0567-2-20

20. Walter K, Bosch S: Intercontinental cross-linking of power supply - calculating an optimal power line corridor from North Africa to Central Europe. Energ Sustain Soc 2013, 3:14. 10.1186/2192-0567-3-14

21. Dooley F, Tyner W, Sinha KC, Quear J, Cox L, Cox M: The impacts of biofuels on transportation and logistics in Indiana. 2009.

22. Ginder R: Potential infrastructure constraints on ethanol production in Iowa. 2006. http://www.econ.iastate.edu/sites/default/files/publications/papers/p3873-2007-07-27.pdf

23. NADO: Ethanol production impacts transportation system. National Association of Development Organizations Research Foundation Transportation Special Report 2007, 2:1–6.

24. AP: Road shuts ethanol plant. Associated Press. 2007. http://www.rapidcityjournal.com/news/state-and-regional/article_4edf0925-4fc1-5106-8d7a-12cadda26bad.html

25. Wakeley HL, Griffin WM, Hendrickson C, Matthews HS: Alternative transportation fuels: distribution infrastructure for hydrogen and ethanol in Iowa. ASCE 2008, 14:262–271.

26. Kissel C, Cassady J: Wind industry promises rural jobs, transportation challenges. 2008. http://66.132.139.69/uploads/nadort020608b.pdf

27. Tidemann M: Turbines for ethanol plant OK'd. Estherville Daily News. 2010. http://www.esthervilledailynews.com/page/content.detail/id/501320.html

28. Tanaka AM, Anastasopoulos PC, Carboneau N, Fricker JD, Habermann JA, Haddock JE: Policy considerations for construction of wind farms and biofuel plant facilities: a guide for local agencies. State Local Gov Rev 2012,44(2):140–149. 10.1177/0160323X12446029

29. Reynolds RE: The current fuel ethanol industry transportation, marketing, distribution, and technical considerations. 2000.

30. African Development Bank Group: Updated environmental and social impact assessment summary: Lake Turkana Wind Power Project. African Development Bank Group, Kenya; 2011.

31. González J, Rodríguez Á, Mora J, Burgos Payán MM, Santos J: Overall design optimization of wind farms. Renewable Energy 2011,36(7):1973–1982. 10.1016/j.renene.2010.10.034

32. Kinoshita T, Ohki T, Yamagata Y: Woody biomass supply potential for thermal power plants in Japan. Appl Energy 2010,87(9):2923–2927. 10.1016/j.apenergy.2009.08.025

33. Kumar A, Cameron JB, Flynn PC: Biomass power cost and optimum plant size in western Canada. Biomass Bioenergy 2003,24(6):445–464. 10.1016/S0961-9534(02)00149-6

34. Ozerdem B, Ozer S, Tosun M: Feasibility study of wind farms: a case study for Izmir, Turkey. J Wind Eng Ind Aerodyn 2006,94(10):725–743. 10.1016/j.jweia.2006.02.004

35. Tensar: Retrieved March 7, 2014, from Wind farm access roads: two decades of floating roads. 2013. http://www.tensar.co.uk/~/media/548441FC4F6F4B6D90EB410B2969585B.ashx

36. The British Wind Energy Association: Best practice guidelines for wind energy development. The British Wind Energy Association, London; 1994.

37. Van Haaren R, Fthenakis V: GIS-based wind farm site selection using spatial multicriteria analysis (SMCA): evaluating the case for New York State. Renew Sustain Energy Rev 2011,15(7):3332–3340. 10.1016/j.rser.2011.04.010

38. McAloon A, Taylor F, Yee W, Ibsen K, Wooley R: Determining the cost of producing ethanol from corn starch and lignocellulosic feedstocks. Technical Report NREL/TP-580–28893. US Department of Agriculture and National Renewable Energy Laboratory, Golden; 2000.

39. Shurson J: Distillers grain by-products in livestock and poultry feeds. 2010. http://www.ddgs.umn.edu/GenInfo/Overview/index.htm

40. AFDC: Data, analysis and trends. Alternative Fuels & Advanced Vehicles Data Center. 2009. http://www.afdc.energy.gov/data/categories/vehicles

41. RFA: Industry Statistics: US fuel ethanol demand. Renewable Fuel Association. 2008. http://www.in.gov/isda/biofuels/

42. Bruglieri M, Liberti L: Optimal running and planning of a biomass-based energy production process. Energy Policy 2008, 36:2430–2438. 10.1016/j.enpol.2008.01.009

43. EIA: Energy and economic impacts of implementing both a 25 percent rps and a 25 percent rfs by 2025. Energy Information Administration, US Department of Energy, Washington, DC; 2007. . Accessed 11 July 2010 http://www.eia.doe.gov/oiaf/servicerpt/eeim/issues.html

44. EIA: Alternative fueling station total counts by state and fuel type. Energy Information Administration, US Department of Energy, Washington, DC; 2009. . Accessed 11 July 2010 http://www.afdc.energy.gov/fuels/stations_counts.html

45. CBEA: The biomass power industry in the United States. California Biomass Energy Aliance. 2003. http://www.calbiomass.org/

46. DNR: Woody biomass feedstock for the bioenergy and bioproducts industries. Indiana Department of Natural Resources. 2010. http://www.extension.purdue.edu/renewable-energy/docs/IBEWG/fo-WoodyBiomass_final.pdf

47. RED: Biomass Milltown Power Plant. Renewable Energy Development. 2009. http://renewableenergydev.com/red/biomass-milltown-power-plant/

48. Bastos CP: Contributions of solar and wind energy to the world electrical energy demand. 2010. http://www.sefidvash.net/fbnr/pdfs/Solar_and_Wind_Energy.pdf

49. Brown LR: Wind energy demand booming: cost dropping below conventional sources marks key milestone in US shift to renewable energy. 2006. http://www.earth-policy.org/index.php?/plan_b_updates/2006/update52

50. EERE: 80-meter wind maps and wind resource potential. 2010. http://www.windpoweringamerica.gov/wind_maps.asp

51. Tchou J: Wind energy in the United States: a spatial-economic analysis of wind power. 2008. http://www.gsd.harvard.edu/academic/fellowships/prizes/gisprize/ay07-08/Jeremy_Tchou.pdf

52. AWEA: US wind energy projects - Indiana. American Wind Energy Association, Washington, DC; 2009.

53. Lautal P, Stewart R, Handler R, Pouryousef H: Michigan Economic Development Corporation Forestry Biofuel Statewide Collaboration Center. Task B1 Evaluation of Michigan Biomass Transportation Systems. Final Report. Michigan Tech Transportation Institute, Rain Transportation Program; 2012. p 112. . Accessed 2 March 2013 http://www.michiganforestbiofuels.org/sites/default/files/Evaluation%20of%20Michigan%20Biomass%20Transportation%20Systems%20-%20FBSCC%20Task%20B1.pdf

54. McGregor B: A reliable waterway system is important to agriculture. 2010. http://www.ams.usda.gov/amsv1.0/getfile?ddocname=stelprdc5083396&acct=atpub

55. Casavant K: Study of rural transportation issues. Technical report, USDA and US-DOT; 2010.

56. Denicof MR: Ethanol transportation backgrounder: expansion of US corn-based ethanol from the agricultural transportation perspective. Technical report, United States Department of Agriculture; 2007.

57. AASHTO: Guide for design of pavement structures. American Association of State Highway and Transportation Officials. Washington DC; 1993.

58. Asphalt Institute (AI): Thickness design—asphalt pavements for highways and streets. 1981.

59. AASHTO: Mechanistic-Empirical Pavement Design Guide. Interim edition: a manual of practice. AASHTO, Washington, DC; 2008:212.

60. Portland Cement Association (PCA): Thickness design for soil-cement pavements. Portland Cement Association 2001, 30:EB068.

61. Asphalt Institute (AI): Thickness design: asphalt pavements for heavy wheel loads. Manual Series No. 23 (MS-23). 2nd edition. Asphalt Institute, Lexington; 2001.

62. Anastasopoulos PC: Performance-based contracting for roadway maintenance operations. Thesis, Purdue University, West Lafayette, Indiana, M.Sc; 2007.

63. Anastasopoulos PC, Labi S, McCullouch BG: Analyzing duration and prolongation of performance-based contracts using hazard-based duration and zero-inflated ran-

dom parameters Poisson models. Transp Res Rec 2009, 2136:11–19. 10.3141/2136-02

64. Anastasopoulos PC, McCullouch BG, Gkritza K, Mannering FL, Sinha KC: Cost savings analysis of performance-based contracts for highway maintenance operations. ASCE Journal of Infrastructure Systems 2010,16(4):251–263. 10.1061/(ASCE)IS.1943-555X.0000012

65. Anastasopoulos PC, Florax RJGM, Labi S, Karlaftis MG: Contracting in highway maintenance and rehabilitation: are spatial effects important? Transp Res Part A: Policy Pract 2010, 44:136–146. 10.1016/j.tra.2009.12.002

66. Anastasopoulos PC, Labi S, McCullouch BG, Karlaftis MG, Moavenzadeh F: Influence of highway project characteristics on contract type selection: empirical assessment. ASCE Journal of Infrastructure Systems 2010,16(4):323–333. 10.1061/(ASCE)IS.1943-555X.0000035

67. Anastasopoulos PC, Islam M, Volovski M, Powell J, Labi S: Comparative evaluation of public-private partnerships in roadway preservation. Transp Res Rec 2011, 2235:9–19. 10.3141/2235-02

68. Anastasopoulos PC, Volovski M, Labi S: Preservation: are 'public private partnerships' cutting costs? Pavement Preservation J 2013,6(3):33–35.

69. Anastasopoulos PC, Haddock JE, Peeta S: Improving system wide sustainability in pavement preservation programming. J Transp Eng-ASCE 2013,140(3):04013012.

70. Wiltsee G, Wiltsee G: Lessons learned from existing biomass power plants. Technical Report NREL/SR-570–26946. National Renewable Energy Laboratory (NREL). US Department of Energy, Washington, DC; 2000.

71. EPA: National Electric Energy Data System. NEEDS. US Environmental Protect Agency. 2006. http://epa.gov/airmarkets/progsregs/epa-ipm/BaseCase2006.html

72. GE: 1.5sl/1.5s Wind turbine. 2004. http://www.ewashtenaw.org/government/departments/planning_environment/planning/wind_power/Monthly%20Data_Reports/Attachment_1.pdf

73. McKendry P: Energy production from biomass. Part 1: overview of biomass. Bioresour Technol 2002, 83:37–46. 10.1016/S0960-8524(01)00118-3

74. Zuo Y, Maness P, Logan BE: Electricity production from steam exploded corn stover biomass. Energy Fuels 2006, 20:1716–1721. 10.1021/ef0600331

75. Mani S, Tabil LG, Sokhansanj S: Grinding performance and physical properties of wheat and barley straws, corn stover and switchgrass. Biomass Bioenergy 2004, 27:339–352. 10.1016/j.biombioe.2004.03.007

76. Armstrong J: Wind farms and county roads. 2009.

77. The Manitowoc Company, Inc: Manitowoc 16000 product guide. Manitowoc, Wisconsin; 2009.

78. The Manitowoc Company, Inc: Manitowoc 999 product guide. Manitowoc, Wisconsin; 2009.

79. Fricker JD, Whitford RK: Fundamentals of transportation engineering: a multimodal approach. Prentice Hall, New Jersey; 2004.

80. Till RD: Overload truck wheel load distribution on bridge decks. Technical Report R-1529. Structural Section, Construction and Technology Division, Michigan Department of Transportation (MDOT), Lansing; 2009.

81. Eamon CD, Nowak AS: LRFD calibration for wood bridges. 2003.

82. Anastasopoulos PC: Infrastructure asset management: a case study on pavement re-habilitation. Ph.D. Dissertation. Purdue University, West Lafayette, Indiana; 2009. Available electronically from . Accessed 15 Aug 2011 http://search.proquest.com/docview/304991168

83. Anastasopoulos PC, Labi S, Karlaftis MG, Mannering FL: Exploratory state-level empirical assessment of pavement performance. ASCE Journal of Infrastructure Systems 2011,17(4):200–215. 10.1061/(ASCE)IS.1943-555X.0000057

84. Anastasopoulos PC, Mannering FL, Haddock JE: Random parameters seemingly unrelated equations approach to the post-rehabilitation performance of pavements. ASCE Journal of Infrastructure Systems 2012,18(3):176–182. 10.1061/(ASCE) IS.1943-555X.0000096

Table 13 is not available in this version of the article. To view this additional information, please use the citation on the first page of this chapter.

CHAPTER 3

Reducing Greenhouse Gas Emissions Through Strategic Management of Highway Pavement Roughness

TING WANG, JOHN HARVEY, AND ALISSA KENDALL

3.1 INTRODUCTION

The national pavement network is a key component of the transportation infrastructure that the modern US economy depends on for mobility and movement of goods. The vehicles that use the network are responsible for nearly a quarter of the US's greenhouse gas (GHG) emissions [1]. In the state of California, on-road vehicle use contributes to an even larger share, comprising about 35% of the state's GHG emissions [2].

In 2006, the California State Legislature passed Assembly Bill 32 (AB 32), the Global Warming Solutions Act, to reduce GHG emissions throughout the state [3]. The California Air Resources Board (CARB), the lead agency for implementing AB 32, estimated the year 2020 baseline

Reducing Greenhouse Gas Emissions Through Strategic Management of Highway Pavement Roughness. © Wang T, Harvey J, and Kendall A. Environmental Research Letters **9**,3 (2014). http://dx.doi.org/10.1088/1748-9326/9/3/034007. Licensed under a Creative Commons Attribution 3.0 Unported License, http://creativecommons.org/licenses/by/3.0/.

emissions at 507 million metric tons (MMT) of CO_2-equivalent (CO_2-e), with 168.1 MMT of CO_2-e from on-road traffic [2].

Although the implementation of AB 32 has led to studies that focus on GHG emission reduction strategies and their cost-effectiveness in many industrial sectors, to date there has been no evaluation of pavement management strategies to help meet its objectives. Pavement management includes the measurement of pavement condition and the programming of maintenance and rehabilitation (M&R) treatments to achieve goals for the pavement network such as maintaining or restoring smoothness, eliminating cracking (which eventually leads to roughness), and improving vehicle fuel economy at minimum cost to the agency and taxpayers.

Numerous studies have demonstrated a life cycle assessment (LCA) approach is needed to provide a comprehensive evaluation of the environmental burdens of a product or process, and to reduce the risk of unintended negative consequences [4]. For pavements, a typical life cycle includes material production, construction, use, M&R, and end-of-life (EOL) phases.

Despite its omission from many previous LCA studies as identified in [4, 5], the pavement use phase is critical to modeling life cycle GHG emissions. Vehicle fuel consumption and emissions are affected by pavement surface characteristics, namely roughness (or "smoothness" from another perspective) as typically measured by the International Roughness Index (IRI), and to a lesser extent by macrotexture as measured by the mean profile depth (MPD) or mean texture depth (MTD). Pavement roughness and macrotexture are the deviations of a pavement surface from a true planar surface with the wavelengths of deviations ranging from 0.5 to 50 m, and from 0.5 to 50 mm, respectively [6, 7]. Roughness characterizes the primary wavelengths that excite shock absorbers in vehicle suspension systems, drive chain components and cause deformation of tire sidewalls for a moving vehicle. Macrotexture influences the tire–road contact patch, which consumes energy through viscoelastic hysteresis of the rubber in the contact patch of a moving tire [6]. Therefore, both wavelengths dissipate energy, lost as waste heat, as a vehicle moves along the pavement. This process is experienced as rolling resistance by vehicles, of which roughness can account for over 80% for a typical California highway.

IRI values can range from about 0.5–5 m km^{-1} (32–315 in mile^{-1}) on a paved high-speed highway, with lower values indicating a smoother surface. The US Federal Highway Administration (FHWA) defines high-speed highway pavements with an IRI greater than 2.7 m km^{-1} (170 in mile^{-1}) as being in 'poor' condition [8], which accounts for about 15% of the pavement network in both the US and California [8, 9].

Because an improvement in smoothness immediately affects every vehicle traveling over the pavement, the cumulative effects on GHG emissions can be substantial in the near term compared to the changes in vehicle technology or land use policy, which may take years to implement. Therefore, this study examines the California pavement network's potential contribution to reducing GHG emissions from on-road traffic through targeted M&R treatments to reduce roughness, by identifying the IRI value that should trigger an M&R event to minimize GHG emissions, as measured by CO_2-e emissions.

The tradeoff on triggering M&R treatment is that if the roughness trigger value is set too low, the materials production and construction processes required to maintain a smooth pavement with frequent M&R treatments can exceed the CO_2-e reduction from improved fuel economy in the use phase. This study also assesses the cost-effectiveness of using the optimized triggers and compares them with other GHG mitigation strategies studied in the existing literature for the transportation sector.

Few, if any, pavement management systems (PMSs) adopted by state transportation agencies have included environmental impacts in their analysis frameworks. However, several academic studies have attempted to integrate pavement management operations with LCA to reduce environmental impacts. These studies, including Lidicker et al [10] and Zhang et al [11], attempted to minimize the environmental impacts in the pavement life cycle for project-level case studies and a very small local road network, respectively, and used relatively simple emission models by optimizing the M&R frequency and intensity through multi-criteria decision analysis.

This letter demonstrates a network-level study of state-owned highways in California that builds on a pavement LCA model described by Wang et al [12]. Although it is implemented on the California network, the approach can be generalized to any pavement network and any set of treatments to assess environmental impacts and support network-level decision-making. This study includes a subset of common M&R treatments used in California for which sufficient information has been observed and collected. The treatments considered are two pavement preservation treatments extensively used in the California Department of Transportation's (Caltrans') Capital Preventive Maintenance (CAPM) program [13]: (1) a medium thickness asphalt overlay applied on all asphalt surfaced pavements, and (2) diamond grinding with slab replacement on concrete surfaced pavement with less than 10% shattered slabs. A rehabilitation treatment, replacement of concrete lanes with new concrete pavement when there are more than 10% shattered slabs, is also included. This last treatment is used far less often than the CAPM treatments.

3.2 METHODS

The pavement network is composed of segments, each of which is described by a set of characteristics that influence the optimal IRI trigger for M&R treatments to reduce GHG emissions, such as traffic volume, traffic composition, and pavement surface condition. Each pavement segment presents a unique combination of these characteristics. Figure 1 shows the analytical approach used in this study, detailed in the following sections.

3.2.1 NETWORK CHARACTERIZATION

Because of the computational and practical complexity of developing thousands of segment-specific triggers, the network is divided into seven groups based on each segment's traffic level as measured by passenger car equivalents (PCEs) [14]. To calculate PCE, each truck is counted as 1.5 equivalent passenger cars regardless of the type of the truck [14]. Traffic level was identified as the most important segment characteristic for determining whether there is a net reduction of CO_2-e emissions from an M&R treatment [12]. Then, the life cycle CO_2-e emissions are calculated for each group over a range of IRI triggers to identify the optimal trigger for reducing CO_2-e emissions for each group. The approach is intended to maintain a balance between computational intensity and thoroughness.

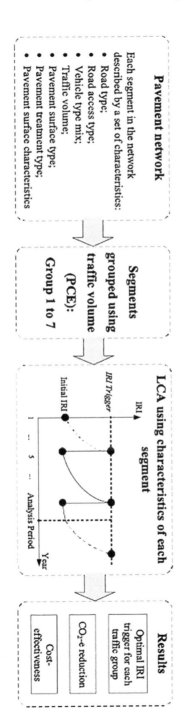

FIGURE 1. Analytical approach of this study.

Table 1 shows the characteristics that describe each segment in the network. A detailed description is included in section S.1 of the supplementary material (available at stacks.iop.org/ERL/9/034007/mmedia).

TABLE 1: Characteristics used to describe each segment in the pavement network.

Characteristic	Pavement life cycle phases involved	Values
Road type	Use	Categorical: rural road; urban road (for consideration of congestion)
Road-access type	Use	Categorical: restricted access (freeway); unrestricted access (highway)
Vehicle type mix	Use	Categorical: passenger cars; 2/3/4/5-axle trucks at Year 2012–2021;
Traffic volume	Use	Continuous numerical: traffic volume of each vehicle type;
Pavement type	Material production, construction, and use	Categorical: asphalt pavement; concrete pavement
Pavement treatment type	Material production, and construction	Categorical: medium asphalt overlay; diamond grinding with slab replacement; concrete lane replacement
Pavement surface characteristics (pavement performance)	Material production, construction, and use	Continuous numerical: IRI performance; MPD performance

Traffic volume can affect CO_2-e emissions in two ways: first, the rate of pavement deterioration, as represented by the performance model of pavement surface characteristics in this study, is affected by the level of truck traffic; and second, pavement roughness affects the fuel economy of every vehicle that uses the pavement, including both passenger cars and trucks. Thus, this study uses the concept of PCE from the Highway Capacity Manual to assist in grouping segments of the network [14]. It should be emphasized that PCE is only used to divide the network into groups. When calculating pavement performance and vehicle fuel economy, segment-specific algorithms for truck traffic (in the form of Equivalent Single Axles Load [ESAL]) and emission factors for each type of vehicle are applied. Traffic volume data is taken from the Caltrans traffic volume report and Caltrans truck traffic database, which reports volume on all the lanes in one direction (termed 'directional segment') [15]. Section S.1 of the supplementary material (available at stacks.iop.org/ERL/9/034007/ mmedia) provides additional information on the data sources and methods of traffic volume.

The cumulative distribution plot of total daily PCE on all directional segments in the network is the basis for grouping the segments into categories, as shown in figure 2. The network is first divided into quartiles, and then to improve calculation of traffic-induced emissions, a finer resolution of 10% intervals is used for those segments above the median. The dividing points are therefore at the 25th, 50th, 60th, 70th, 80th and 90th percentiles in the plot, which correspond to total daily PCEs on directional segments of 2517, 11 704, 19 108, 33 908, 64 656, and 95 184, respectively.

Pavement surface characteristics, i.e., the metrics that represent pavement performance in this study, are modeled with explanatory variables of treatment type, truck traffic (in terms of ESAL) and climate region. Each type of pavement treatment and surface characteristic has a specific performance model with different formats and inputs. For asphalt overlay, IRI is a function of initial IRI, ESAL, climate region and pavement age, whereas MPD is a determined by a number of variables such as asphalt mix type and truck traffic. For concrete pavement, IRI after diamond grinding with slab replacement is a function of initial IRI and cumulative ESAL, whereas MPD is a function of pavement age and climate region. These performance models are detailed in section S.2 in the supplementary material (available at stacks.iop.org/ERL/9/034007/mmedia). During the characterization the network, the performance of each segment is modeled using the applicable pavement, truck traffic and climate region properties of this segment that can affect the performance. If a property also contributes to other parts of the life cycle modeling, such as total vehicle traffic, it is then listed separately as a characteristic in table 1.

3.2.2 LIFE CYCLE ASSESSMENT

This study performs life cycle GHG calculations on each pavement segment and sums the results within each traffic group. The GHGs tracked in this study include carbon dioxide (CO_2), methane (CH_4) and nitrous oxide (N_2O) and they are normalized to CO_2-e using the IPCC's 100-year global warming potentials [16]. The scope of the analysis includes material production, construction, and use phases. Only the transport of materials removed during the treatments is modeled for the EOL phase. This study mainly focuses on repeated treatments with relatively short design lives, so a 10-year analysis period (2012–2021) is adopted to cover approximately 1.5 times the design lives.

In the life cycle modeling, each directional segment in the network is evaluated through two scenarios: (1) the M&R scenario and (2) the Do Nothing scenario. Then, the results are compared to current and historical Caltrans policies for IRI triggers. Caltrans historically used an IRI trigger of 3.54 m km^{-1} (224 in mile^{-1}) for asphalt pavement and 3.36 m km^{-1} (213 in mile−1) for concrete pavement [23]. Recently, Caltrans has changed to a trigger of 2.86 m km^{-1} (170 in mile^{-1}) for all pavements. These policies are, in practice, constrained by budget limitations, meaning that pavement roughness often exceeds trigger values until funding is sufficient.

In the M&R scenario, when the IRI of a segment reaches the trigger, a treatment is performed, bringing down the IRI based on historical Caltrans data. The emissions and cost from the material production and construction of the treatment are calculated based on the material quantity and construction activity. The use phase CO_2-e is calculated based on the pavement surface characteristics and traffic composition and volume. The well-to-wheel (WTW) emissions of fuels are always used when there is fuel consumption.

In the Do Nothing scenario, the pavement is maintained at approximately its current roughness and macrotexture using repairs by local Caltrans forces. Emissions from material production and construction for these localized repairs are not calculated due to uncertainty in the particular activities and materials that might be used, and the fact that only small quantities of material are likely to be used. The use phase emissions for the Do Nothing scenario are calculated similarly to the M&R scenario. It should be noted that the state would never implement a Do Nothing strategy on the entire network, and would only implement a Do Nothing strategy on those sections where they do not have sufficient funding, with the constrained funding resulting in a de facto implementation of a Do Nothing strategy.

The difference in CO_2-e emissions between these two scenarios is calculated over the analysis period. This procedure is repeated for all segments in the network and the difference from each segment is summed for the final result over the analysis period. Ten IRI triggers, evenly distributed from 0.4 to 4.4 m km^{-1} (38–279 in mile^{-1}) are assessed for each traffic group and the value that leads to the highest CO_2-e reduction is considered optimal. The selection of the IRI triggers is intended to cover the common range of IRI values on modern paved highways in the US. It should be emphasized that the "optimal triggers" developed in this study only apply to the CO_2-e emission reduction on the modeled highway using the selected maintenance treatments. Other social benefits such as increased safety, and social disbenefits such as diversion of funding for other purposes, are not included in the analysis and the results may not be optimal considering a broader range of objectives or a larger system definition.

3.2.3 COST-EFFECTIVENESS

Cost-effectiveness describes the cost of abatement per unit of pollution (here metric tons of CO_2-e emission, or tCO_2-e). A lower cost-effectiveness value indicates less money is needed to achieve the same level of CO_2-e reduction. This study assesses two types of costs: agency cost and modified total cost. Agency cost reflects the total contracted expenditures of the transportation agency, while the modified total cost is the agency cost subtracting the cost of saved fuel for the road users. A negative modified total cost indicates that this measure in the long term can reduce CO_2-e as well as save money for the two stakeholders considered (agency and road users) and is therefore a "no-regret" strategy. A total cost calculation would consider additional costs of rougher pavement due to vehicle maintenance, vehicle life, accidents, etc. However, high-quality data for these costs are not readily available, which is why a modified total cost is used.

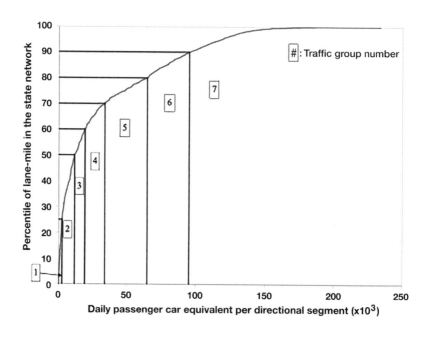

FIGURE 2: Cumulative distribution plot of daily PCE per directional segment and traffic group.

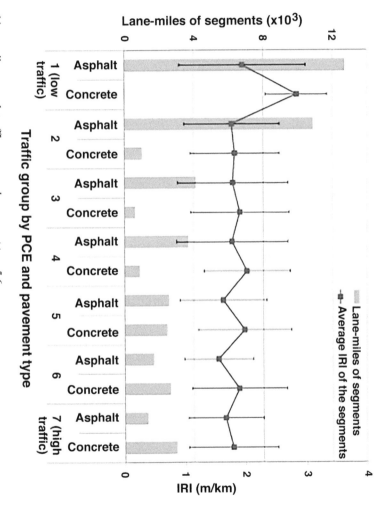

FIGURE 3: IRI and lane-miles on each traffic group and pavement type 5,6.

3.3 INPUT DATA

3.3.1 STATE PAVEMENT NETWORK

Figure 3 shows some descriptive statistics of the highway network based on the traffic groupings. Pavement type and IRI data are from the 2011 Caltrans Automated Pavement Condition Survey, and used as the initial state for the analysis in this study. Overall, asphalt surfaced pavement accounts for about 76% of the total lane-miles, mostly the segments in Groups 1–4.

3.3.2 LCA MODEL

3.3.2.1 MATERIAL AND CONSTRUCTION PHASES

The modeling of emissions from the material production and construction phases is described in the project-level study that this network-level study builds on [12]. When applied to the network, the modeling of these phases is calculated based on the materials quantities and total lane-miles of each treatment.

For cost analysis, the agency cost of each treatment is acquired from the Caltrans PMS [17]. The fuel price for the saved energy consumption is acquired from the US Annual Energy Outlook [18]. A discount rate of 4% is used in accordance with Caltrans' practice for life cycle cost analysis [19]. Detailed life cycle emissions data and cost information are included in section S.3 of the supplementary material (available at stacks.iop.org/ERL/9/034007/mmedia).

The selection and timing of treatments roughly follow Caltrans guidelines [13] and the decision tree in the Caltrans PMS for the treatments modeled in this study [17], with the assumption that pavement surface type (asphalt or concrete) does not change. Detailed design of these treatments can be found in section S.1 of the supplementary material (available at stacks.iop.org/ERL/9/034007/mmedia).

The effect from work zone traffic, either through additional fuel use and traffic delay cost from congestion, or fuel savings caused by vehicles operating at slightly slower speeds in the work zone, is not considered in this study because construction of the modeled treatments on high traffic segments will generally be performed at night, causing almost no traffic delay and therefore minimal impacts. On the other hand, major rehabilitation or reconstruction treatments, although not modeled in this study, often occur during the daytime and can cause substantial traffic congestion. The cost and emission from their work zone congestion should be included in the modeling.

TABLE 2: Factorial variables used to develop vehicle tailpipe CO2 emission factors.

Pavement type	Road type	Road-access type	Vehicle type mix	Pavement surface characteristics
Asphalt pavement; Concrete pavement	Urban roads; Rural roads	Restricted-access road; Unrestricted-access road	Passenger cars; 2-axle truck; 3-axle truck; 4-axle truck; 5 or more axle truck, including fuel efficiency improvement from 2012 to 2021	IRI performance; MPD performance
Categorical variable	Categorical variable	Categorical variable	Categorical variable	Continuous variable

3.3.2.2 USE PHASE

The use phase of the pavement life cycle considered in this study includes the additional CO_2-e from vehicle operation due to pavement deterioration. Because CO_2 contributes over 99.8% of the vehicle tailpipe CO_2-e emissions, other tailpipe GHG emissions are not included. The well-to-pump (WTP) CO_2-e emissions for fuel are included based on vehicle fuel consumption using the GREET model [20].

To conduct the network-level analysis, vehicle tailpipe CO_2 emission factors are developed as a function of selected pavement segment characteristics, as shown in table 2. Sensitivity analyses were performed to evaluate whether additional characteristics were needed to represent the network's heterogeneity. The characteristics include the effects of congestion on urban restricted-access roads and different road vertical gradients on mountainous roads. Both had very small impacts on the relationship between pavement roughness and fuel consumption, and therefore were omitted.

The vehicle tailpipe CO_2 emission factors were developed as a continuous function of MPD and IRI for each combination of the categorical variables. A series of IRI and MPD values under each combination of the categorical variables were modeled using MOVES to calculate the tailpipe CO_2 emission [21], and then linear regression was used on the results to develop the function. The R-squared of the regression is above 0.99 in all cases, indicating that the vehicle tailpipe CO_2 emission factor is highly linearly correlated with IRI and MPD for each combination of the categorical variables. Section S.5 of the supplementary material (available at stacks.iop.org/ERL/9/034007/mmedia) provides additional details on these calculations.

Because pavement surface characteristics are inputs in the use phase and they change every year, the performance models for IRI and MPD developed by Tseng [22], Lu et al [23] and Rao et al [24] are used. These models are mainly functions of truck traffic level and climate, detailed in section S.2 of the supplementary material (available at stacks.iop.org/ERL/9/034007/mmedia).

3.4 RESULTS AND DISCUSSION

3.4.1 COMPARISON OF M&R WITH DO NOTHING

Figure 4 shows the annualized CO_2-e reduction when the modeled treatments are performed using different IRI triggers. The x-axis shows the IRI value that triggers the modeled treatments, and the y-axis shows the annualized CO_2-e emissions reduction compared to Do Nothing over the 10-year analysis period under different triggers. A positive value means there is a net reduction of CO_2-e.

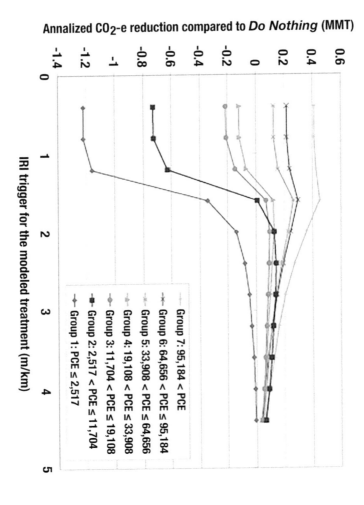

FIGURE 4: Annualized CO_2-e reductions versus IRI trigger for different traffic level over the 10-year analysis period for entire state network compared to Do Nothing.

Figure 4 shows that the higher the traffic level, the lower the IRI trigger that results in the maximum life cycle CO_2-e emissions reduction. Table 3 shows the maximum emission reductions in each group, the corresponding IRI triggers, and the modified total cost-effectiveness. Detailed cost-effectiveness results are in table S7 of the supplementary material (available at stacks.iop.org/ERL/9/034007/mmedia).

The ten percent of the network with the highest traffic (Group 7) yields nearly 35% of the CO_2-e emissions reductions, despite similar or lower roughness compared to the next lower traffic groups. For the segments that make up the bottom quartile of the network based on traffic volume (daily PCE lower than 2517) there is no IRI trigger that yields a reduction, indicating emissions from the material production and construction phases are always higher than reductions during the use phase.

TABLE 3: IRI trigger for the maximum CO_2-e reductions compared to Do Nothing over the 10-year analysis period for the entire network.

Traffic group	Daily PCE	Total lane-miles	Percentile of lane-miles	Optimal IRI trigger in m km^{-1a}	Annual-ized CO_2-e reductions (MMT)	Modified total cost-effectiveness[b] ($/tCO$_2$-e)
1	<2517	12 068	0–25	—	0	N/A
2	2517–11 704	12 068	25–50	2.4 (152)	0.141	1169
3	11 704–19 108	4 827	50–60	2.0 (127)	0.096	857
4	19 108–33 908	4 827	60–70	2.0 (127)	0.128	503
5	33 908–64 656	4 827	70–80	1.6 (101)	0.264	516
6	64 656–95 184	4 827	80–90	1.6 (101)	0.297	259
7	>95 184	4 827	90–100	1.6 (101)	0.45	104
Total					1.38	416

[a]*inch mile^{-1} is in the parentheses. "Optimal" here only applies to CO_2-e reductions and does not include other social benefits.*

[b]*N/A = not applicable since no net CO_2-e reduction. "Modified total cost" is the agency cost subtracting the cost of saved fuel for the road users. Agency cost, while not shown here, is the total contracted expenditures of the transportation agency. Detailed cost-effectiveness results are in table S7 of the supplementary material (available at stacks.iop. org/ERL/9/034007/mmedia).*

The annualized CO_2-e emissions reduction that can be achieved if these optimal IRI triggers are implemented is 1.38 MMT over 10 years compared to Do Nothing. For comparison, CARB has estimated that the average annual baseline emissions from on-road vehicles is about 168.1 MMT CO2-e between 2006 and 2020 [2]. Therefore, the potential reduction estimated from this study would contribute to about a 0.8% decrease compared to Do Nothing.

The IRI triggers for the maximum CO_2-e reductions are not the same as those which lead to the highest modified total cost-effectiveness (detailed results shown in table S8 of the supplementary material (available at stacks.iop.org/ERL/9/034007/mmedia)). In fact, in all traffic groups, the highest cost-effectiveness occurs at the IRI trigger of 4.4 m km^{-1}; the highest trigger assessed. This is because the relative change in cost with a higher IRI trigger is always greater than the relative change in CO_2-e emissions. In a given analysis period, a higher IRI trigger leads to fewer treatments and thus lower agency cost, but brings a relatively larger drop of IRI and thus greater reduction in emissions. As a result, as the IRI trigger increases, the cost-effectiveness increases. However, this conclusion may change if longer life treatments and total road user costs are evaluated.

The results in table 3 show that CO_2-e reductions from performing the modeled treatments on rough pavements has the potential to contribute to the statewide GHG reduction target, and that the traffic level plays an important role in determining optimal triggers. The cost-effectiveness provides a guide for prioritizing projects: the segments with a high cost-effectiveness, such as Group 7, should receive a higher priority for treatments when under a budget constraint.

3.4.2 COMPARISON WITH CALTRANS' IRI TRIGGERS

Caltrans' PMS prioritization policies prior to 2011 used an IRI trigger of 3.54 m km^{-1} (224 in $mile^{-1}$) for asphalt pavement and 3.36 m km^{-1} (213 in $mile^{-1}$) for concrete pavement [25]. Since 2011, the trigger has been 2.86 m km^{-1} (170 in $mile^{-1}$) for all pavements. In practice, meeting these policy goals is constrained by budget, which does not permit all segments in the network to receive planned treatments.

By interpolating this study's results, the historical and current Caltrans IRI triggers lead to an annualized CO_2-e reduction of 0.57 and 0.82 MMT compared to Do Nothing over 10 years, with a modified total cost-effectiveness of \$355/tCO_2-e and \$520/tCO_2-e, respectively. Therefore, compared to the historical trigger, the current trigger of 2.86 m km^{-1} substantially reduces CO_2-e, although it is less cost-effective. The complete results of using these triggers are shown in table S9 and S10 of the supplementary material (available at stacks.iop.org/ERL/9/034007/mmedia). Compared to the historical and current Caltrans IRI triggers, the optimal IRI triggers can achieve an annualized marginal CO_2-e reduction of 0.82 and 0.57 MMT, with a marginal modified total cost-effectiveness of \$457/tCO_2-e and \$266/tCO_2-e, respectively. The current Caltrans IRI trigger is much closer to the optimal IRI triggers than the historical triggers, and this leads to a very small marginal cost change and an improved cost-effectiveness. The complete results of the comparison are in table S11 and S12 of the supplementary material (available at stacks.iop.org/ERL/9/034007/mmedia).

In practice, even if the IRI of a segment has reached its designated trigger, a treatment may not occur until 1–3 years later because of project development and delivery time, or longer if there are budget constraints. Therefore, the actual CO_2-e reductions and the cost in the analysis period are likely to be reduced.

Table 4 shows how the CO_2-e reduction and cost can change if the M&R activity is delayed. For a two-lane (per direction) 1-mile rural freeway with a one-direction annual average daily traffic of 12 000 and 10% trucks (PCE of 12 600), the treatment should be triggered at 2 m km^{-1} (127 in $mile^{-1}$). If the treatment is performed 1, 2, or 3 years after the IRI reaches the trigger, the CO_2-e reductions can drop by approximately 6%, 13%, and 18%, respectively, compared to on-time treatment. Also evident is that the cost drops faster than the CO_2-e reductions. Although the delay can lead to better cost-effectiveness, in part because fewer treatments are triggered in the analysis period, it reduces the potential CO_2-e reductions.

TABLE 4: Example of comparison between on-time and late triggering (10-year analysis period).

When is treatment performed	Total agency cost compared to Do Nothing (dollar)	Annualized CO_2-e reduction compared to Do Nothing (metric ton)	Agency cost ratio (compared to on-time treatment)	CO_2-e reduction ratio (compared to on-time treatment)
On time	8.72×10^4	6.22×10^4	1.00	1.00
1 year later	7.90×10^4	5.85×10^4	0.91	0.94
2 years later	7.16×10^4	5.39×10^4	0.82	0.87
3 years later	7.04×10^4	5.08×10^4	0.81	0.82

3.4.3 COMPARISON WITH ALTERNATIVE GHG MITIGATION MEASURES

Lutsey examined GHG mitigation strategies for the transportation sector and their cost-effectiveness [26]. The cost-effectiveness of the pavement management treatments in this study are considerably lower than many alternative measures Lutsey identified, which were as low as $60/tCO2-e or less, as shown in table 5 [26].

This result for pavement occurs because the construction of civil infrastructure is expensive, and more importantly, the costs evaluated in this study only include the agency and fuel cost, and exclude other road user costs. Because the main functionality of pavement is to maintain the mobility of goods and people using vehicles, one of the primary purposes for pavement management is to ensure the safety and efficiency for transportation, which is what a road user cares most about. Therefore, a more comprehensive benefit analysis would include other social benefits such as vehicle life, safety, tire consumption, goods damage, vehicle maintenance, driver comfort, and the value of time. From this point of view, the CO_2-e reduction can be considered a "co-benefit" from pavement management when used as a GHG mitigation measure, and will be more cost-effective if all road user costs are included.

A preliminary study showed that while the fuel consumption (and therefore fuel cost) exhibits a linear relationship with roughness, the total road user cost can increase exponentially with the pavement roughness [27]. The ratio between total road user cost and fuel cost ranges from 6 to 10, depending on the vehicle type, driving speed and pavement condition [27]. A first-order estimate shows that the total cost-effectiveness can range from −$710/tCO$_2$-e to −$1610/tCO$_2$-e (compared to the $416/tCO$_2$-e as shown in table 3) if all road user costs are included. This result indicates that pavement management, when properly programmed like in this study, can potentially be a cost-competitive measure to reduce GHG emissions if total road user cost is considered. Once the total cost models as a function of pavement roughness for California are fully developed, the comparison with other transportation strategies should be performed again.

TABLE 5: Comparison of cost-effectiveness between pavement and some alternative measures in the transportation sector [26].

Measure	Annual CO$_2$-e emission reduction[a]	Total life cycle cost-effectiveness ($2008/tCO$_2$-e)[b]
Light duty vehicle: incremental efficiency	20% tailpipe reduction	−75
Light duty vehicle: advanced hybrid vehicle	38% tailpipe reduction on new vehicles	42
Commercial trucks: class 2b efficiency	25% tailpipe reduction	−108
Alternative refrigerant	Replacement of HFC-134a with R-744a (CO2)	67
Ethanol fuel substitution	Increase mix of cellulosic ethanol to 13% by volume	31
Biodiesel fuel substitution	Increase mix of biodiesel to 5% by volume	51
Aircraft efficiency	35% reduction in energy intensity	−9
Strategic pavement roughness triggers (this study)	1.38 MMT	390

[a]*The first seven measures show the value of CO2-e emission reduction in 2025. The value from "strategic pavement roughness triggers (this study)" is an annualized life cycle value between 2012 and 2021.*
[b]*The result from table 3 is in 2012 dollars and is converted to 2008 dollars in this table using consumer price index (CPI).*

3.4.4 UNCERTAINTY AND SENSITIVITY ANALYSES

The main input data for this study include the traffic count and IRI on the state pavement network, the emission factors from the MOVES model, maintenance cost and IRI performance.

The traffic count used in this study is extracted from the traffic database used by the Caltrans PMS. It incorporates the high-quality data from Caltrans Performance Measurement System (PeMS) and Weigh-In-Motion stations. The IRI on the network was collected in the 2011 Caltrans Automated Pavement Condition Survey. Because of their wide use within Caltrans, these two sources of data have gone through a number of quality control and quality assurance studies to ensure their accuracy and should have minimal uncertainty. For emission factors, because MOVES itself does not provide an uncertainty analysis module, it is very difficult to perform any uncertainty analyses outside this complex model. Further, because this study is focused on the emission difference between scenarios, the uncertainty of emission factors can be expected to play a less important role. For maintenance cost, although it is averaged from historical Caltrans construction projects and there are some uncertainties associated with it, it can be predicted that the impact on the result is completely linear because this study does not include cost in the optimization procedure.

Therefore, sensitivity analyses are performed on two variables to assess their impacts on the results: constructed smoothness and analysis period. Complete results can be found in table S13 and S14 of the supplementary material (available at stacks.iop.org/ERL/9/034007/mmedia).

For constructed smoothness, three levels of initial IRI after construction are considered based on the statistical analysis described in section S.2.1 of the supplementary material (available at stacks.iop.org/ERL/9/034007/mmedia). The results show that the constructed smoothness can change the optimal triggers by as much as 0.8 m km^{-1}(51 in mile^{-1}). With a good construction smoothness, the benefit from the treatment can be more than doubled compared to the average construction smoothness; likewise, with poor constructed smoothness the benefits can be reduced by more than half. The constructed smoothness is primarily controlled by construction practice, quality control and the existing pavement condition, and to a lesser degree by the treatment type. Some "Best Practices" to improve the constructed smoothness include pre-paving/grinding planning and preparation, good mix design, grade control, equipment control and good communication between personnel [28, 29]. Construction smoothness has historically not been specified in terms of IRI in California and most other states due to technical difficulties; a specification based on a moving beam has been used to identify "bumps" which were then removed before acceptance of the completed project. However, those difficulties have recently been solved and many states are now moving to specification of construction in terms of IRI [30]. The new specifications are expected to reduce average IRIs obtained from treatment as well as variability. For example, California implemented an IRI based construction smoothness specification in July, 2013. However, data are not yet available to analyze the marginal benefit from this and other specific practices to improve smoothness.

Sensitivity analysis on the analysis period was performed to assess whether the selection of a particular time horizon substantially influences the results, using three analysis periods: 10, 15, and 20 years. The results show that different selections do not substantially change the optimal IRI triggers. One explanation for the small effect of the analysis period is that all periods selected covered the design life of most treatments, and this study amortizes the impact from material production and construction phases of the last treatment to avoid horizon effects.

3.5 CONCLUSIONS AND FUTURE WORK

In this letter, a pavement LCA model is applied to the California state pavement network to evaluate the CO_2-e reduction resulting from a strategic application of selected M&R treatments. This approach and methodology can be adapted to any road network by substituting the appropriate local or regional conditions, treatments and practices, and used to support network-level pavement decision-making.

In this study, the network is broken into different groups based on their traffic level. An optimal IRI trigger leading to the highest CO_2-e reduction is developed for each group. These IRI triggers are only optimized for CO_2-e reduction for this study and may not lead to a socially optimal result. The following conclusions are drawn from this study:

Traffic level has a substantial impact on the optimal IRI trigger. With optimal triggering, annualized CO_2-e reductions of 1.38, 0.82, and 0.57 MMT can be achieved compared to the Do Nothing, the historical, and the current Caltrans IRI triggers over the 10-year analysis period. The cost-effectiveness of these CO_2-e reduction strategies is worse than those reported for other transportation sector CO_2-e abatement measures when only considering fuel cost savings, but preliminary analyses indicate that pavement management can potentially be a cost-competitive measure to reduce GHG emissions if total road user cost is considered.

Delaying M&R treatment when the IRI has reached the designated trigger can considerably reduce potential CO_2-e reduction.

Sensitivity analyses show that the constructed smoothness has a substantial impact on the results, and the analysis period does not have a substantial impact on the optimal IRI triggers. The potential for changes in cost-effectiveness of treatment in light of recently improved construction smoothness specifications warrants future investigation.

Future implementation of this work will include the expansion of the treatment options using an approach similar to this study, such as major rehabilitation/reconstruction treatments. An upcoming study will investigate the impact on fuel consumption from pavement structure change. With this expansion of scope, it is possible to develop a more comprehensive M&R schedule and policy to reduce CO_2-e emissions over the pavement network life cycle.

REFERENCES

1. US Environmental Protection Agency 2013 Inventory of US Greenhouse Gas Emissions and Sinks: 1990–2011 (Washington, DC: US Environmental Protection Agency)
2. California Air Resources Board 2011 Greenhouse Gas Inventory—2020 Emissions Forecast (Sacramento, CA: California Air Resources Board)
3. California Air Resources Board 2008 Climate Change Proposed Scoping Plan: A Framework for Change (Sacramento, CA: California Air Resources Board) 133
4. Santero N J, Masanet E and Horvath A 2011 Life-cycle assessment of pavements. Part I: critical review Resour. Conserv. Recy. 55 801–9

5. Kendall A 2007 Concrete infrastructure sustainability: life cycle metrics, materials design, and optimized distribution of cement production School of Natural Resources and Environment and Civil and Environmental Engineering (Ann Arbor, MI: University of Michigan) 176

6. Sandberg U and Ejsmont J A 2002 Tyre/Road Noise Reference Book (Kisa, Sweden: InformEx)

7. Sayers M W 1995 On the calculation of international roughness index from longitudinal road profile Transp. Res. Rec. 1501 1–12

8. US Department of Transportation, Federal Highway Administration and Federal Transit Administration 2010 2010 Status of the Nation's Highways, Bridges, and Transit: Conditions and Performance (Washington, DC: US Department of Transportation)

9. Caltrans 2011 2011 State of the Pavement Report (Sacramento, CA: Office of Roadway Maintenance, Division of Maintenance, California Department of Transportation)

10. Lidicker J, Sathaye N, Madanat S and Horvath A 2013 Pavement resurfacing policy for minimization of life-cycle costs and greenhouse gas emissions J. Infrastruct. Syst. 19 129–37

11. Zhang H, Keoleian G A, Lepech M D and Kendall A 2010 Life-cycle optimization of pavement overlay systems J. Infrastruct. Syst. 16 310–22

12. Wang T, Lee I S, Kendall A, Harvey J, Lee E B and Kim C 2012 Life cycle energy consumption and GHG emission from pavement rehabilitation with different rolling resistance J. Cleaner Prod. 33 86–96

13. Caltrans 2011 Capital Preventive Maintenance (CAPM) Guidelines (Sacramento, CA: California Department of Transportation) 8

14. Transportation Research Board 2010 Highway Capacity Manual (Washington, DC: Transportation Research Board)

15. Caltrans 2011 Traffic Data Branch (Sacramento, CA: California Department of Transportation)

16. IPCC 2007 IPCC fourth assessment report: climate change 2007 (AR4): the physical science basis Contribution of Working Group I to the Fourth Assessment Report of the IPCC ed Solomon et al (Cambridge: Cambridge University Press)

17. Harvey J, Lea J, Tseng E, Kim C and Kwan C 2012 PaveM Engineering Configuration (Davis, CA: University of California Pavement Research Center)

18. US Energy Information Administration 2013 Annual Energy Outlook 2013 (Washington, DC: US Energy Information Administration) 233

19. Caltrans 2010 Life Cycle Cost Analysis Procedures Manual (Sacramento, CA: California Department of Transportation) 143

20. Argonne National Laboratory 2011 GREET Model: The Greenhouse Gases, Regulated Emissions, and Energy Use in Transportation Model (Chicago, IL: Argonne National Laboratory)

21. US EPA 2010 MOVES2010a User Guide (Washington, DC: US Environmental Protection Agency)

22. Tseng E 2012 The construction of pavement performance models for the california department of transportation new pavement management system Department of Civil and Environmental Engineering (Davis, CA: University of California) 94

23. Lu Q, Kohler E, Harvey J T and Ongel A 2009 Investigation of Noise and Durability Performance Trends for Asphaltic Pavement Surface Types: Three-Year Results (Davis, CA: University of California Pavement Research Center) 206

24. Rao S, Yu H T, Khazanovich L, Darter M I and Mack J W 1999 Longevity of diamond-ground concrete pavements Transp. Res. Rec. 1684 128–36

25. Caltrans 1997 Maintenance and Rehabilitation Priority Assignment Based on Condition Survey (Sacramento, CA: California Department of Transportation) 20

26. Lutsey N 2008 Prioritizing climate change mitigation alternatives: comparing transportation technologies to options in other sectors Institute of Transportation Studies (Davis, CA: University of California) 197

27. The World Bank 2010 HDM-4 Road Use Costs Model Version 2.00 (Washington, DC: The World Bank)

28. FHWA 2002 HMA Pavement Smoothness: Characteristics and Best Practices for Construction (Washington, DC: Federal Highway Administration)

29. FHWA 2002 PCC Pavement Smoothness: Characteristics and Best Practices for Construction (Washington, DC: Federal Highway Administration)

30. Karamihas S M, Gillespie T D, Perera R W and Kohn S D 1999 Guidelines for Longitudinal Pavement Profile Measurement (Washington, DC: Transportation Research Board)

There are several supplemental files that are not available in this version of the article. To view this additional information, please use the citation on the first page of this chapter.

PART II

URBAN TRANSPORTATION

CHAPTER 4

Infrastructure and Automobile Shifts: Positioning Transit to Reduce Life-Cycle Environmental Impacts for Urban Sustainability Goals

MIKHAIL CHESTER, STEPHANIE PINCETL, ZOE ELIZABETH, WILLIAM EISENSTEIN, AND JUAN MATUTE

4.1 BACKGROUND

It is widely accepted that the combustion from passenger vehicle tailpipes is a leading cause of environmental pollution and emerging life-cycle approaches present an opportunity to better understand how transit investments reduce transportation impacts. In California, automobile travel is responsible for 38% of statewide greenhouse gas (GHG) emissions and other pollutants have been linked to significant health impacts [1, 2]. California's Assembly Bill 32 calls for the reduction of statewide GHG emissions to 1990 levels by 2020. To achieve this, a suite of strategies will be deployed, including Senate Bill 375 which requires regional transportation

Infrastructure and Automobile Shifts: Positioning Transit to Reduce Life-Cycle Environmental Impacts for Urban Sustainability Goals. © *Chester M, Pincetl S, Elizabeth Z, Eisenstein W, and Matute J. Environmental Research Letters* **8** *(2013), http://dx.doi.org/10.1088/1748-9326/8/1/015041. Licensed under a Creative Commons Attribution 3.0 Unported License, http://creativecommons.org/licenses/by/3.0/.*

plans to achieve GHG emissions targets from the transportation system and may induce cities to deploy new public transit systems.

Passenger vehicles, however, do not exist in isolation as they require a large and complex system to support a vehicle's operation. To understand the environmental impacts from transportation systems, and more importantly how to cost-effectively minimize these impacts, it is necessary to include vehicle, infrastructure, and energy production life-cycle components, in addition to operation [3]. A life-cycle approach is particularly important for new mass transit systems that produce large upfront impacts during the deployment of new infrastructure systems for long-run benefits in the reduction of automobile travel [4]. However, little is known about the life-cycle environmental benefits and costs of deploying public transit systems to meet urban energy and environmental goals. Using the city of Los Angeles as a case study, a life-cycle assessment (LCA) of the Orange bus rapid transit (BRT) and Gold light rail transit (LRT) lines is developed. These transit lines, both deployed in the past decade, provide an opportunity to better understand how new transit systems will help cities reduce transportation impacts.

4.2 METHODOLOGY

An environmental LCA is developed for the Orange BRT, Gold LRT, and competing automobile trips. The LCA includes vehicle (e.g., manufacturing and maintenance), infrastructure (e.g., construction and operation), and energy production components, in addition to vehicle propulsion effects [3]. To inform a broad array of transportation policy and decision makers, two different LCA framings are used: attributional and consequential. The attributional framing evaluates the long-run average footprint of each system allocating impacts to a passenger-mile-traveled (PMT). It includes, for example, the construction impacts of the existing road system for an automobile trip. However, given the importance of understanding how public transit investments contribute to urban sustainability goals, a consequential analysis of the decision to build each system is also produced, culminating in a cumulative impact savings at some future time. The consequential analysis answers how the BRT and LRT systems may contribute to Los Angeles (LA) meeting its Senate Bill 375 GHG and air quality goals. The results from the attributional and consequential approaches should be considered independently.

4.2.1 LIFE-CYCLE CHARACTERISTICS OF LOS ANGELES TRANSPORTATION SYSTEMS

The LCA methods used follow those reported in existing research by the authors, however, significant efforts were made to obtain system-specific data from LA Metro [5] and model life-cycle impacts with regionalized energy mixes and processes. Extensive details are provided in Chester et al [6] and the following discussion focuses on the data collection and methods used to assess the dominating life-cycle processes. For each mode, near-term (at maturity, in the 2020–2030 time period) and long-term (2030–2050) vehicles are modeled.

4.2.1.1 ORANGE LINE BRT

The Orange BRT is an 18 mile dedicated right-of-way running east–west through the San Fernando Valley. The line opened in 2005 and now services 25 500 riders per day, exceeding initial forecasts [7]. The line is viewed as a tremendous success; service has been increased to meet the latent demand, its construction has induced 140 000 new annual bike trips on the buffering green belt, and has roused development [8–10]. The initial line consists of a two-lane asphalt roadway connecting 18 stations, sometimes buffered by landscaping. In 2012 a 4 mile extension from Canoga Park to Chatsworth was opened. Orange BRT buses are 60 foot compressed natural gas (CNG) articulated North American Bus Industry vehicles with the structure, chassis, and suspension (54% of weight) manufactured in Hungary and final assembly occurring in Alabama [5]. Vehicle manufacturing is assessed with Ecoinvent's bus manufacturing processes using current and projected European mixes [11–13]. The buses use conventional lead-acid batteries with an expected lifetime of 13 months [5]. The energy and emissions effects from ocean going vessel transport (Hungary to Alabama) and driving the buses from Alabama to LA are included [14]. LA Metro expects buses to last 15 years [5]. Engineering design documents are used to determine busway characteristics. The western-most segment of the line uses local roadways and the 17 mile dedicated busway consists of roughly 13 miles of asphalt and 4 miles of concrete surface layers. Recycled materials were used for the subbase. Asphalt wearing layers, concrete wearing layers, and the subbase are modeled with PaLATE [15] and are assumed to have 20, 15, and 100 year lifetimes. Stations are also included and are designed as a raised concrete platform [16]. The construction and maintenance of the 4700 parking spaces are also assessed with PaLATE. The Orange BRT line uses 1.2 GWh of electricity [5] purchased from LA Department of Water and Power (LADWP) for infrastructure operation including roadway, station, and parking lot lighting, and is evaluated with GREET [14]. Routine maintenance of vehicles and infrastructure are modeled with SimaPro [13].

Orange BRT vehicle operation effects are based on emissions testing by the California Air Resources Board (CARB) of similar bus engines [17, 18]. CARB results for urban duty drive cycles are used and assume that buses will use three way catalysts in the near-term. The emission profiles are validated against other testing reports for similar vehicles and engines [19–28]. In the long-term, it is assumed that Orange BRT buses will achieve fuel economies consistent with best available technology buses today (effectively a 23% improvement from today's buses) and that the CARB 2020 certification standards are met which require 75–85% reductions in air pollutants [17]. The extraction, processing, transport, and distribution of CNG for the buses are evaluated with GREET [14] including upstream effects.

4.2.1.2 GOLD LINE LRT

The Gold line is an expanding rail system that extends from downtown LA to east LA and Pasadena, with plans to triple the line length in the coming decades. The system began operation in 2003 and currently consists of 19.7 mile of at-grade, retained fill, open cut, and aerial sections. LA Metro uses 54 tonne AnsaldoBreda P2550 2-car 76-seat trains manufactured in Italy and shipped by ocean going vessel to LA. Train manufacturing was assessed with SimaPro [13] with current and future European electricity mixes [12] and transport with GREET [14]. The infrastructure assessment is based on engineering design documents [29] which are used to develop a material and construction equipment assessment following the methods used by Chester and Horvath (2009) [4]. The unique construction activities associated with track sections are assessed and detailed characteristics are reported in Chester et al (2012) [6]. There are currently 21 stations of which 19 are at-grade. Satellite imagery is used to determine the area of station platforms which are designed as steel-reinforced concrete slabs on a subbase. The Gold line has 2300 parking spaces across 9 stations, and these are assessed with PaLATE [15]. Electricity consumption data were provided to the research team by LA Metro and are from meters at stations and maintenance yards [5]. In 2010, 20 GWh were purchased from LADWP, 3.2 GWh from Pasadena Water and Power, and 1.2 GWh from Southern California Edison, and propulsion electricity use accounts for roughly one-half of the total [30]. Given the dominating share of LADWP electricity consumed, the utility is used to assess the air emissions of electricity production [14]. Currently, 39% of LADWP electricity is produced from coal and there are plans to phase this fuel out by 2030 as the utility transitions their portfolio towards renewable targets [31]. The 2030 LADWP mix will use more natural gas and renewables and would decrease electricity generation GHG emissions by 50% and SOx by 60% [14]. Vehicle and infrastructure maintenance and insurance impacts are also modeled [6].

Gold line trains consume approximately 10 kWh of electricity per vehicle mile traveled (VMT) [30] and current and future electricity mixes are assessed to determine near-term and long-term vehicle footprints. The 2030 LADWP mix is used for long-term train operation where the generation of propulsion electricity produces fewer GHG and CAP emissions. Primary fuel extraction, processing, and transport to the generation facility (i.e., energy production) effects are modeled with GREET [14].

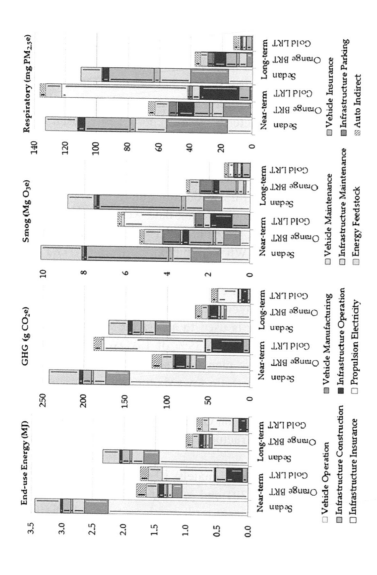

FIGURE 1: Life-cycle per PMT results for average occupancy vehicles. For each impact both near-term and long-term results are shown for each mode. Vehicle tailpipe effects are gray, vehicle are blue, infrastructure are red, and energy production are green. Local impacts are shown with a line on the left of the life-cycle result and remote on the right if the dominating share of effects occurs inside or outside of LA county.

4.2.1.3 ORANGE BRT AND GOLD LRT I NDIRECT AUTOMOBILE EFFECTS

The Orange BRT and Gold LRT lines produce indirect automobile effects through new station access and egress by auto travel. Additionally, the Orange BRT's new biking and walking infrastructure avoids auto trips. The cumulative effect is included in the LCA. 7% of transit riders drive alone to the station and 3% from the stations [9]. These trips are between 1.7 and 2.5 miles [32]. LA Metro [8] estimates that the Orange BRT's biking and walking shift reduces auto annual VMT between 71 000 and 540 000. The indirect auto effects of transit implementation are included in the life-cycle footprint of the Orange BRT and Gold LRT lines, averaged over all PMT.

4.2.1.4 COMPETING AUTOMOBILE TRIP

While LA has an extensive and well-utilized public transportation network, the large sprawling region is dominated by automobile travel at 85% of trips (or 97% of PMT), biking and walking at 13%, and transit at 2% [32]. New transit lines have experienced success in reducing automobile travelers, with (in 2009) 25% of Orange BRT passengers having previously made the trip by auto and 67% of Gold line travelers [9, 33]. Consequently, the assessment of the Orange BRT and Gold LRT lines should consider the life-cycle effects of competing automobile trips to assess the traveler's environmental footprint had transit not existed. The avoided automobile effects are also necessary for evaluating the net change of air pollutants in the region as a result of new transit options.

An automobile trip that substitutes an Orange BRT or Gold LRT line trip is assessed. The transit lines are expected to operate indefinitely so representative automobiles are selected to assess near- (35 mile gallon^{-1}, 3000 lb) and long-term (54 mile gallon^{-1}, 1800 lb) car travel [14]. The long-term automobile is modeled with a lighter weight to assess technology changes that may be implemented to meet aggressive fuel economy standards. Both automobiles are estimated to have a 160 000 mile lifetime. A transport distance of 2000 mile from the manufacturing plant to LA is included by class 8b truck. Infrastructure construction is based on a typical LA arterial segment allocated by annual VMT facilitated [34], and modeled with PaLATE [15]. Vehicle insurance and infrastructure construction and maintenance are also included [6].

The near and long-term automobiles are modeled with 35 and 54 mile gallon^{-1} standards in GREET [14] to assess emerging fuel economy standards in the long life expectancy of the new transit systems. Petroleum extraction, processing, and transport effects assuming California Reformulated Gasoline and 16% oil sands are modeled.

4.2.1.5 RIDERSHIP AND MODE SHIFTS

Orange BRT and Gold LRT ridership have been steadily increasing since the lines opened and forecasts for future ridership are developed to assess long-term effects. From its first year of operation to 2009, the Orange BRT has increased yearly boardings from 6.1 to 8.4 million, and the Gold line from 4.8 to 7.6 million [7]. This corresponds to 49 and 55 million PMT in 2009 for the respective systems, an increase of 30% (in 5 years) and 36% (in 7 years). Future ridership estimates are developed using 2035 station access forecasts developed by LA Metro. A polynomial interpolation is used to assess adoption between now and 2035 when an estimated 100 and 130 million annual PMT are delivered by the respective systems [5]. In 2009 the Orange BRT average occupancy was 37 with 57 seats and the Gold line 43 with 72 seats per car [7]. The average occupancy of automobile travel in LA is 1.7 passengers for all trips, 1.4 for households that also use transit, and 1.1 for work trips [32]. Auto trip purpose characteristics are joined with transit onboard survey results and future forecasts to determine avoided automobile travel. Currently, 25% of Orange BRT and 67% of Gold LRT previous trip takers would have made the trip by automobile [9, 33]. Given that fuel prices are expected to increase and the transit lines are expanding to auto dominated regions, may be interconnected with other transit lines [35], and are anticipated to experience further development [10], auto shift forecasts are developed to 2050. Using future trip and station access forecasts from LA Metro, the current auto shift growth rates are extrapolated resulting in a median long-term shift of 52% for the

Orange BRT and 80% for the Gold LRT. Furthermore, to assess avoided automobile travel distance from transit shifts, a clustering approach was used to determine that across household income, workers, and vehicles, one PMT shifted to the Orange BRT or Gold LRT lines avoids one PMT of automobile travel [32, 36].

4.2.2 ENERGY AND ENVIRONMENTAL INDICATORS AND STRESSORS

An energy and environmental life-cycle inventory is developed and then joined with photochemical smog formation and human health respiratory impact stressors. The inventory includes end-use energy and emissions of greenhouse gases (GHGs), NOx,SOx, CO, PM_{10}, $PM_{2.5}$, and VOCs. GHGs are reported as CO_2-equivalence (CO_2e) for a 100 year horizon using radiative forcing multipliers of 25 for CH_4 and 298 for N_2O. Los Angeles has struggled to meet National Ambient Air Quality Standards for PM and ozone so inventory results are joined with impact characterization factors from the Tool for the Reduction and Assessment of Chemical and Other Environmental Impacts (TRACI, v2) to assess respiratory and smog stressors [37]. A stressor is the upper limit of impacts that could occur and not the actual impact that will occur. The deployment of these new transit systems may help LA reduce GHG emissions to meet environmental goals. However, by assessing a broad suite of environmental indicators, unintended tradeoffs (i.e., reducing one impact but increasing another) can be identified early and mitigation strategies developed.

4.3 MODAL PASSENGER MILE COMPARISONS

The Orange BRT and Gold LRT lines will reduce life-cycle per PMT energy use, GHG emissions, and the potential for smog formation at the anticipated near-term and long-term ridership levels. However, given the $PM_{2.5}$ intensity of coal-fired electricity generation powering the Gold LRT, there is a potential for increasing out-of-basin respiratory impacts in the near-term, highlighting the unintended tradeoffs that may occur with disconnected GHG and air quality policies. The Gold LRT respiratory impact potential is the result of coal electricity generation in LADWP's mix and associated mining activities. The coal-fired Navajo Generating Station (NGS) in Arizona and the Intermountain Power Plant (IPP) in Utah are owned, at least in part, by LADWP and the utility is planning to divest in the plants by 2025 [31]. The NGS and IPP are two of the largest coal-fired power plants in the Western US and have been targeted for emissions reductions, primarily to improve visibility at nearby parks including the Grand Canyon [38, 39]. However, secondary particle formation from NOx and SOx, in addition to $PM_{2.5}$, have been shown to be a respiratory concern despite the each facility's remote location [39–41]. LADWP is aggressively pursuing divestiture in its 21% share of NGS and 100% share of IPP which will lead to significant long-term benefits for the Gold LRT [31].

Figure 1 shows that significant environmental benefits can be achieved by automobiles, Orange BRT, and Gold LRT in the long-term as a result of established energy and environmental policies as well as vehicle technology changes, and that public transit technology and energy changes will produce more environmental benefits per trip than automobiles. In the near-term, both the Orange BRT and Gold LRT lines can be expected to achieve lower energy and GHG impacts per PMT than emerging 35 mile gallon^{-1} automobiles. While propulsion effects (vehicle operation and propulsion electricity) constitute a majority share of life-cycle effects for energy and GHGs, vehicle manufacturing, energy production, and in the case of the Gold line, electricity for infrastructure operation (train control, lighting, stations, etc) contribute significantly. Due to high NOx and $PM_{2.5}$ emissions in coal-fired electricity generation, Gold LRT in the near-term creates large potential smog and respiratory impacts, however, the replacement of this coal electricity with natural gas by 2015–2025 will result in significant reductions in the long-term [31]. For non-GHG air emissions, indirect and supply chain processes (in this case vehicle manufacturing and infrastructure construction) typically dominate the life-cycle footprint of modes showing how vast supply chains that traverse geopolitical boundaries result in remote impacts far from where the decision to build and operate a transportation mode occurs. Diesel equipment use, material processing, and electricity generation for the production and distribution of materials throughout the supply chain generate heavy NOx and $PM_{2.5}$

emissions that when allocated to LA travel can dominate the life-cycle smog and respiratory effects.

In the long-term, automobile fuel economy gains, reduced emission buses, and RPS electricity will have the greatest impacts on passenger transportation energy use and GHG emissions in LA. Larger renewable shares feeding Orange BRT bus manufacturing drives a 46% reduction in life-cycle respiratory impacts. For the Gold LRT, RPS electricity will reduce both propulsion and infrastructure operation smog effects by 93%. Automobile indirect effects show non-negligible contributions to life-cycle impacts when allocated across all trip takers. The impacts of station access and egress by motorized travel are explored in later sections.

LCA transcends geopolitical boundaries in its assessment of indirect and supply chain processes, and urban sustainability policy makers should recognize that local vehicle travel triggers energy use and emissions outside of cities. This is clear for coal-fired electricity generation in Arizona but can become complex when moving up the supply chain for vehicle and infrastructure components. Vehicle operation and propulsion electricity effects are a large portion of energy consumption and GHG emissions and will occur locally while energy production (i.e., primary fuel extraction and processing) and vehicle manufacturing occur remotely. For the sedan, roughly 72–77% of life-cycle energy consumption and GHG emissions occurs locally meaning that for every 75 MJ of energy consumed or grams of CO_2e emitted in LA, an additional 25 are triggered outside of the city. For the Orange BRT, local energy use and GHG emissions constitute 74–82% and for the Gold LRT, only 53–62% due to electricity generation both outside of the county and the state. These percentages change significantly for smog and respiratory stressors due to the larger contributions of non-propulsion effects in the life-cycle.

For the sedan, remote electricity generation for vehicle manufacturing and energy production emissions mean that only 52–73% of potential impacts may occur locally. Similarly, remote vehicle manufacturing and CNG production emissions for the Orange BRT result in roughly 55–76% of respiratory impact stressors occurring locally. Due to out-of-state coal electricity generation, in the near-term the Gold LRT line has the lowest fraction of life-cycle smog and respiratory effects occurring locally, at 54% and 31%. Urban energy and environmental goals should recognize that cities rely on complex and dynamic energy and material supply chain networks and that it may be possible through contracts or supplier selection to reduce remote impacts. This will only occur if policymakers adopt an environmental assessment framework that acknowledges that cities are not isolated systems and trigger resource use and emissions that exist beyond their geopolitical boundaries [42].

The per PMT assessment is valuable for understanding how regions should allocate their total emissions or impacts to each mode's travel and identify which life-cycle processes should be targeted for the greatest environmental gains, however, a consequential assessment is needed for assessing how new modes will contribute to a city reaching their environmental goals, by comparing against a regional baseline.

4.4 PUBLIC TRANSIT FOR ENERGY
AND ENVIRONMENTAL GOALS

To assess the effects of the decision to deploy a public transit system and how such a system contributes to a city reaching an environmental goal, a consequential LCA framework must be used. The decision to deploy the Orange and Gold lines resulted in the operation of new vehicles that require infrastructure and trigger life-cycle processes that consume energy and generate emissions. While induced demand is created, reduced automobile travel has also occurred [9, 33], which should reduce future energy consumption and emissions from personal vehicles. A consequential LCA is used to assess the increased impacts from new transit modes and avoided impacts of reduced automobile travel. Future adoption forecasts [5] are used with mode shift survey results to develop the decadal benefit–cost impact assessment and payback estimates shown in figure 2.

For both transit lines, construction impacts (light red life-cycle bar) begin the series in the first decade. Starting in the second decade the transit systems begin operation, offsetting automobile travel, and over the coming decades reach ridership maturity. For both modes and all impacts, the benefits from reduced automobile travel outweigh the environmental costs of the new transit systems. The avoided impacts are 1.5–3 times larger for GHG emissions than the added transit emissions, 1.3–5.5 times for smog, and 1.4–15 times for respiratory impacts. There are significantly fewer impacts produced from the initial construction of the dedicated Orange BRT right-of-way than from the Gold LRT tracks due to a variety of process, material, and supply chain life-cycle effects. The heavy use of concrete for Gold line tracks results in significant CO_2, VOC, and $PM_{2.5}$ releases during cement and concrete production due to calcination of limestone and emissions of organics elements and fine particles during kiln firing. The result is that the Orange line payback for GHGs and respiratory effects is almost immediate and the Gold line paybacks occur 30–60 years after operation begins. The results highlight the sensitivity of payback to auto trips shifted to transit. Transportation planners can position new transit to help cities meet environmental goals by developing strategies that ensure certain levels of automobile shifting are achieved to accelerate paybacks. Figure 3 shows the payback speed for energy consumption and air emissions as a response to the percentage of transit trip takers that have shifted from automobiles.

Figure 3 shows that with greater shifts to transit from automobiles paybacks occur more quickly. This response will hold true for any public transit system where the per PMT effects of automobiles are larger than the public transit mode. The life-cycle energy and emissions curves have different intercepts and trajectories showing that paybacks will occur at different rates, and will not be the same for any environmental indicator. A transit system that achieves 50% of riders having shifted from automobiles will experience a different payback date for VOCs than it will for GHGs. The Orange line's 2009 25% shift is currently below the 30% needed to produce energy and CO reductions, however, given the anticipated full adoption shift of 52% the system will over 50 years produce a net reduction of 320 Gg CO_2e (see figure 2). The information in figure 3, while specific to the LA transit systems, can provide valuable goals for cities. For any mode, there is a minimum window of percentage of transit riders shifted from automobiles where pollutants will be reduced. For example, the abscissa for the Orange BRT reveals that at between roughly 10% and 30% pollutant reductions will be achieved (how quickly is a separate question). At 30%, the Orange BRT system is guaranteed to have payback across all pollutants. This maximum in the window can be used by cities that are discussing the implementation of new transit systems to help meet environmental goals. Planning efforts should be coordinated such that the systems achieve these minimum mode shifts.

FIGURE 2: Environmental impact schedules and resulting paybacks. Decadal (D) life-cycle results (bars) are shown for the new transit system (red) and avoided automobile (blue) effects. Cumulative (i.e., net effects) life-cycle and local green lines are shown and when they cross the abscissae have resulted in a net reduction of impacts as a result of the transit system. When payback occurs, the net benefits are shown at the bottom of the decade.

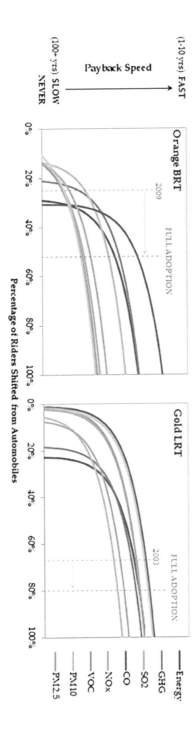

FIGURE 3: Transit energy and environmental payback speed with automobile shifts. The different payback speed curves are shown as a function of the percentage of transit riders having shifted from automobiles. For each mode, current and forecasted full adoption levels are shown as dashed vertical gray lines. At the abscissa, payback does not occur.

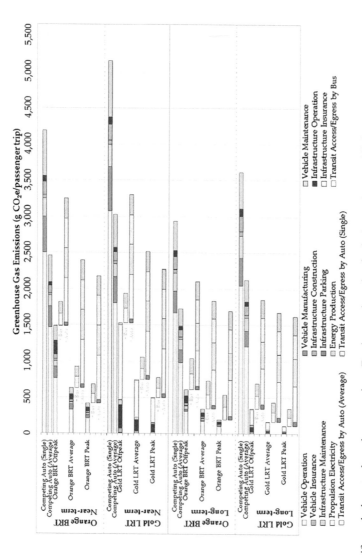

FIGURE 4: Life-cycle door-to-door ghg comparison. For the transit modes, feeder bus and automobile impacts should be assessed cumulatively. For example, an Orange BRT or Gold LRT trip that starts or ends with a bus trip is equal to the transit life-cycle bar plus the bus purple striped bars. Feeder bus and auto travel results are shown with both operational (striped bar with white background) and life-cycle (striped bar with gray background) portions. For auto access/egress to transit, the effects of average occupancy (1.7) passengers per car is shown as orange stripes and if the automobile was single occupancy then the red striped bar would be added on. Bike and walk impacts are not considered.

4.5 DOOR-TO-DOOR LIFE-CYCLE EFFECTS

Transportation environmental policies should consider the multi-modal door-to-door impacts of trips, and LCA can provide valuable insight for both the operational and non-operational effects of a traveler's choice. A life-cycle understanding of door-to-door trips is particularly important for transit travelers whose access or egress to stations occurs by automobile where questions arise of the benefits of these trips, particularly when infrastructure (specifically station parking) is included. A door-to-door GHG LCA is developed for a unimodal automobile trip compared against each transit line. For each transit line, access/egress is shown with local bus service as well as by automobile. Typical trip distances are used (as described in previous sections) and processing of LA Metro and travel survey data provides information on feeder bus and automobile typical trip characteristics [7, 8, 32, 33]. The typical Orange BRT trip is 6 mile with feeder bus and auto trips adding on average 1.8 and 4.2 mile respectively. A competing unimodal auto trip is 10.2 mile assuming distance shifts identified in the previously described mode shift clustering analysis [7, 32]. The typical Gold LRT trip is 7.5 mile with feeder bus and auto trips of 3.3 and 5 mile and is compared against a 12.5 mile competing auto trip [7, 32, 33]. Current and future Orange BRT, Gold LRT, and feeder bus offpeak, average, and peak occupancies are determined from LA Metro data and forecasts [5, 7]. Auto feeder travel is shown as both average and single occupancy travel. The impacts of the 4700 and 2300 parking spaces (shown as bright orange in figure 1) are now shifted to the automobile feeder trips. The results are shown in figure 4 in both the near-term and long-term for GHG emissions for offpeak, average, and peak travel.

The Orange BRT and Gold LRT door-to-door trips with typical access/egress by other local buses or automobiles are likely to have a lower life-cycle footprint than a competing unimodal automobile trip. The only exceptions are offpeak (low occupancy) transit travel with single occupancy automobile feeder access/egress compared against average (1.7 passenger) unimodal auto trips. Transit travel (even with single occupancy automobile feeder access/egress) consistently produces lower impacts than a competing single occupancy automobile trip. On average, transit+local bus trips have 77% lower GHG trip footprints than a competing automobile trip and transit+auto 52% lower. Recent onboard travel surveys report that 49% of Orange BRT passengers arrive to or leave from stations by local transit and 14% by automobile, and for the Gold line 41% link bus and other rail trips (data on access/egress by automobile were not identified) [9, 33]. Strategies that shift travelers from automobiles to public transit-only service produces the greatest environmental benefits and parking infrastructure management is central to changing behavior [43]. Figure 4 shows that parking construction and maintenance (orange bar) impacts for transit+auto trips can be as large as the transit infrastructure construction and maintenance (pink bars) per trip. These infrastructure enable the emergent travel behavior and the provision of low cost or free parking at LA Metro stations helps to encourage the auto access/egress impacts [43]. Environmental benefit–cost analyses should consider the life-cycle tradeoffs of land use for station parking versus transit-oriented development (TOD) and the co-benefits that could be achieved by TODs in reducing both auto access/egress impacts and household energy use [44].

4.6 INTEGRATING TRANSPORTATION LCA IN URBAN ENVIRONMENTAL POLICYMAKING

Public transit systems are typically positioned as transportation environmental impact reducers and as policy and decision makers begin to incorporate life-cycle thinking into planning, new strategies must be developed for integrating LCA. The results show that both local and remote life-cycle environmental impacts will be reduced by implementing BRT and LRT for all impacts in the long-term. The results also show that the decision to implement a new transit system in a city has significant local and remote energy and environmental impacts beyond vehicle operation. These life-cycle impacts are the result of indirect and supply chain processes that are often ignored by policy and decision makers, as well as environmental mitigation strategies.

Challenges exist for implementing life-cycle results in governmental processes [45]. Because life-cycle emissions are distributed across numerous air basins throughout the United States and the world, there exists a spatial mismatch for policymaking. Both transportation planning and emissions control policy structures in the United States are fragmented across jurisdictions and across different components of the life-cycle. The transportation system is created through a series of federal, state, regional and local programs and authorities acting in an independent, yet interdependent, manner. Designing a policy structure to reduce life-cycle emissions is therefore a complex task, and a variety of policy options may be viable. While there is no simple policy fix, mitigation strategies that effectively incorporate LCA into transportation planning should involve all of the following:

1. analytical and decision criteria for project selection;
2. improving the capability to compare different transportation modes to one another in planning and project financing processes;
3. improving the capability to conduct analysis of complex environmental impacts into transportation planning before project selection occurs (i.e. not only in post-decisional environmental impact assessments);
4. improving analytical integration across different spatial and temporal scales; and,
5. creating purchasing strategies that emphasize the use of products and materials with higher recycled content and establish relationships with suppliers that have instituted efficiency measures.

Given these needs, the metropolitan region is likely the most useful geographic scale for transportation LCA integration and LCA can be used as a valuable guiding framework for novel mitigation strategies. Metropolitan Planning Organizations (MPOs) already offer the greatest planning integration across modes and already possess relatively advanced analytical and planning capabilities for the development of Regional Transportation Plans. Pigovian tax or cap-and-trade structures for carbon or other emissions can use life-cycle results to capture indirect and supply chain impacts and if cast at a large geographic scale can reduce urban and hinterland impacts by transcending the notion that activities in cities are contained within a geopolitical boundary.

REFERENCES

1. CARB 2011 California Greenhouse Gas Emissions Inventory: 2000–2009 (Sacramento, CA: California Air Resources Board)
2. Ostro B et al 2007 The effects of components of fine particulate air pollution on mortality in California: results from CALFINE Environ. Health Perspect. 115 13–9
3. Chester M and Horvath A 2009 Environmental assessment of passenger transportation should include infrastructure and supply chains Environ. Res. Lett. 4 024008
4. Chester M and Horvath A 2012 High-speed rail with emerging automobiles and aircraft can reduce environmental impacts in California's future Environ. Res. Lett. 7 034012
5. LA Metro 2012 Personal Communications with Los Angeles County Metropolitan Transportation Authority: Emmanuel Liban (2011–2012), John Drayton (July 15, 2011), Alvin Kusumoto (August 2, 2011), Scott Page (August 2, 2011), James Jimenez (July 26, 2011), and Susan Phifer (Planning Manager, August 3, 2011) (Los Angeles, CA: Los Angeles County Metropolitan Transportation Authority)
6. Chester M et al 2012 Environmental Life-Cycle Assessment of Los Angeles Metro's Orange Bus Rapid Transit and Gold Light Rail Transit Lines (Arizona State University Report No. SSEBE-CESEM-2012-WPS-003) (Tempe, AZ: Arizona State University)
7. LA Metro 2011 Bus and Rail Ridership Estimates and Passenger Overview (Los Angeles, CA: Los Angeles Metropolitan Transportation Authority)
8. LA Metro 2011 Metro Orange Line Mode Shift Study and Greenhouse Gas Emissions Analysis (Los Angeles, CA: Los Angeles Metropolitan Transportation Authority)
9. Flynn J et al 2011 Metro Orange Line BRT Project Evaluation (Washington, DC: Federal Transit Administration)
10. City of LA 2012 Final Program Environmental Impact Report for the Warner Center Regional Core Comprehensive Specific Plan (Los Angeles, CA: City of Los Angeles)
11. US Energy Information Administration 2012 Annual Energy Outlook (Washington, DC: US Department of Energy)
12. EEA 2010 National Renewable Energy Action Plan Data from Member States (Copenhagen: European Environment Agency)
13. SimaPro 2012 SimaPro v7.3.3 Using the Ecoinvent v2.2 Database (Amersfoort: PRé Consultants)
14. GREET 2012 Greenhouse Gases, Regulated Emissions, and Energy Use in Transportation (GREET; 1: Fuel Cycle; 2: Vehicle Cycle) Model (Argonne, IL: Argonne National Laboratory)
15. PaLATE 2004 Pavement Life-cycle Assessment for Environmental and Economic Effects (Berkeley, CA: University of California)
16. LA Metro 2000 San Fernando Valley East–West Transit Corridor, Major Investment Study (Los Angeles, CA: Los Angeles Metropolitan Transportation Authority)
17. CARB 2000 Risk Reduction Plan to Reduce Particulate Matter Emissions from Diesel-Fueled Engines and Vehicles (Sacramento, CA: California Air Resources Board)
18. Gautam M et al 2011 Testing of Volatile and Nonvolatile Emissions for Advanced Technology Natural Gas Vehicles (Morgantown, WV: West Virginia University)

19. Ayala A et al 2003 Oxidation catalyst effect on CNG transit bus emissions Technical Report (Warrendale, PA: Society of Automotive Engineers) doi:10.4271/2003-01-1900

20. Ayala A et al 2002 Diesel and CNG heavy-duty transit bus emissions over multiple driving schedules: regulated pollutants and project overview Technical Report (Warrendale, PA: Society of Automotive Engineers) doi:10.4271/2002-01-1722

21. NREL 2006 Washington Metropolitan Area Transit Authority: Compressed Natural Gas Transit Bus Evaluation (Golden, CO: National Renewable Energy Laboratory)

22. NREL 2005 Emission Testing of Washington Metropolitan Area Transit Authority (WMATA) Natural Gas and Diesel Transit Buses (Golden, CO: National Renewable Energy Laboratory)

23. Kado N et al 2005 Emissions of toxic pollutants from compressed natural gas and low sulfur diesel-fueled heavy-duty transit buses tested over multiple driving cycles Environ. Sci. Technol. 39 7638–49

24. Lanni T et al 2003 Performance and emissions evaluation of compressed natural gas and clean diesel buses at New York city's Metropolitan transit authority Technical Report (Warrendale, PA: Society of Automotive Engineers) doi:10.4271/2003-01-0300

25. Clark N et al 1999 Diesel and CNG transit bus emissions characterizaion by two chassis dynamometer laboratories: results and issues Technical Report (Warrendale, PA: Society of Automotive Engineers) doi:10.4271/1999-01-1469

26. Nylund N-O et al 2004 Transit Bus Emission Study: Comparison of Emissions from Diesel and Natural Gas Buses (Finland: VTT)

27. Ayala A et al 2003 CNG and diesel transit bus emissions in review 9th Diesel Engine Emissions Reduction Conference (Newport, RI: US EPA)

28. [28ICCT 2009 CNG Bus Emissions Roadmap: From Euro III to Euro VI (Washington, DC: The International Council on Clean Transportation)

29. LACTC 1988 Draft Environmental Impact Report for the Pasadena–Los Angeles Rail Transit Project (Los Angeles, CA: Los Angeles County Transportation Commission)

30. USDOT 2009 National Transit Database (Washington, DC: US Department of Transportation)

31. LADWP 2011 Power Integrated Resource Plan (Los Angeles, CA: Los Angeles Department of Water and Power)

32. NHTS 2009 National Household Travel Survey (Oak Ridge, TN: US Department of Transportation's Oak Ridge National Laboratory)

33. LA Metro 2004 Gold Line Corridor Before/After Study Combined Report (Los Angeles, CA: Los Angeles Metropolitan Transportation Authority)

34. USDOT 2012 National Transportation Statistics (Washington, DC: US Department of Transportation)

35. LA Metro 2012 Sepulveda Pass Corridor Systems Planning Study (Los Angeles, CA: Los Angeles County Metropolitan Transportation Authority)

36. Fraley C and Raftery A E 2002 Model-based clustering, discriminant analysis, and density estimation J. Am. Stat. Assoc. 97 611–31

37. Bare J C 2002 TRACI: the tool for the reduction and assessment of chemical and other environmental impacts J. Indust. Ecol. 6 49–78

38. EPA 2013 Joint Federal Agency Statement Regarding Navajo Generating Station (Washington, DC: US Environmental Protection Agency)

39. GAO 2012 Air Emissions and Electricity Generation at US Power Plants (Washington, DC: US Government Accountability Office)

40. Eatough D J et al 1996 Apportionment of sulfur oxides at canyonlands during the winter of 1990—I. Study design and particulate chemical composition Atmos. Environ. 30 269–81

41. Wilson J C and McMurry P H 1981 Studies of aerosol formation in power plant plumes—II. Secondary aerosol formation in the Navajo generating station plume Atmos. Environ. 15 2329–39

42. Chester M, Pincetl S and Allenby B 2012 Avoiding unintended tradeoffs by integrating life-cycle impact assessment with urban metabolism Curr. Opin. Environ. Sustain. 4 451–7

43. Shoup D 2011 The High Cost of Free Parking (Chicago, IL: American Planning Association)

44. Kimball M et al 2012 Policy Brief: Transit-Oriented Development Infill in Phoenix can Reduce Future Transportation and Land Use Life-Cycle Environmental Impacts (Arizona State University Report No. SSEBE-CESEM-2012-RPR-002) (Tempe, AZ: Arizona State University)

45. Eisenstein W, Chester M and Pincetl S 2013 Policy options for incorporating life-cycle environmental assessment into transportation planning Transp. Res. Rec. at press

CHAPTER 5

The Use of Regional Advance Mitigation Planning (RAMP) to Integrate Transportation Infrastructure Impacts with Sustainability: A Perspective from the USA

JAMES H. THORNE, PATRICK R. HUBER, ELIZABETH O'DONOGHUE, AND MARIA J. SANTOS

5.1 INTRODUCTION

Over half of all humans live in cities, (UNPF 2007) with 1.75 billion more people expected by 2030 (McDonald et al 2011). Urbanization associated with the growth of cities transforms natural landscapes and impacts biodiversity, ecosystem processes and agriculture (Theobald et al 2000, Schwartz et al 2006, Grimm et al 2008a, 2008b, McKinney 2008, Satterthwaite et al 2010). Expanding transportation infrastructure is linked to the functionality of these expanding urban areas (Barthelemy et al 2013,

The Use of Regional Advance Mitigation Planning (RAMP) to Integrate Transportation Infrastructure Impacts with Sustainability: A Perspective from the USA. © Thorne JH, Huber PR, O'Donoghue E, and Santos MJ. Environmental Research Letters *9*,6 (2014). http://dx.doi.org/10.1088/1748-9326/9/6/065001. Licensed under a Creative Commons Attribution 3.0 Unported License, http://creativecommons.org/licenses/by/3.0/.

Schneider and Mertes 2014). However, transportation structures have adverse impacts to the natural environment (Trombulak and Frissell 2000, Forman et al 2003, National Research Council 2005), including: direct and cumulative mortality to species hit by vehicles (Seo et al 2013), reduced dispersal capacity (Forman and Alexander 1998), impediments to gene flow (Epps et al 2005), the spread of invasive species (Gelbard and Belnap 2003), the generation of greenhouse gas emissions (Fuglesvedt et al 2008, Kennedy et al 2009), and landscape fragmentation (Jaeger et al 2005, Girvetz et al 2008). Concern over the rapid growth of cities and transportation infrastructure highlights the need to reduce environmental impacts associated with this growth (Thorne et al 2005, 2006, 2009a) by initiating or improving requirements to offset those impacts via conservation or restoration of other lands, here termed environmental mitigation.

Finding a balance between new transportation infrastructure and preservation of land for biodiversity conservation, ecosystem processes, agriculture, and other needs is most effectively assessed and implemented using planning principles a regional level (Kark et al 2009, Huber et al 2010, Gordon et al 2013, Moilanen et al 2013). We used the first part of a framework called Regional Advance Mitigation Planning (RAMP, figure 1), that uses a regional planning approach to estimate the spatially cumulative impacts from multiple projects. RAMP is part of emerging practice for transportation officials in local, state, and federal government in many areas in the United States (Brown 2006, Marcucci and Jordan 2013). Using the San Francisco Bay Area as a study area, we asked, 'Can early regional assessment of impacts from multiple future transportation projects create added benefits for conservation?' We assessed potential impacts to species, habitats, agriculture and streams from 181 planned road and railroad projects. We show that the early quantification of the impacts, done in aggregate as opposed to the typical project-by-project impact assessment approach can provide the foundation for more ecologically effective and cost effective environmental mitigation strategies. We assert that the methods of the RAMP framework can be applied to many other urbanizing regions of the world.

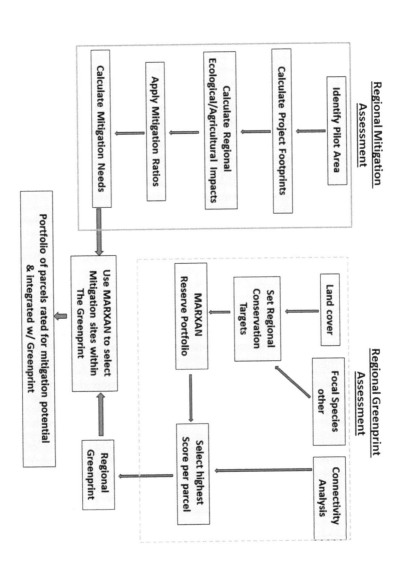

FIGURE 1: The RAMP framework. This framework allows the integration of environmental impact assessment and environmental mitigation (blue box) with regional conservation concerns (dotted box).

5.2 BACKGROUND

RAMP can be thought of as a set of guiding policies, a framework, and a method. The policy component of RAMP emerges when government bodies make conscious decisions to use the framework to address impacts from regional transportation (or other infrastructure) development. Motivations to adopt the framework are available because RAMP addresses several known inefficiencies in the way transportation and other infrastructure projects have typically been implemented in the United States. Inefficiencies that local, state and federal government transportation agencies have historically experienced include: project-by-project environmental mitigation, usually toward the end of a project's timeline, which exposes the agencies to increased costs due to real estate appreciation, and also to losing potential land acquisition opportunities to encroaching development and speculation (Thorne et al 2006). In addition, project-by-project mitigation often overlooks regional and ecosystem scale impacts to species, habitats and ecosystem services, a recognized need in environmental impact assessment (EIA) approaches (Jay et al 2007). Further motivation arises from the possibility that bundling environmental mitigation from several projects in RAMP may permit the acquisition of fewer properties (with larger area) with consequent lower transaction and long-term land management costs, and by potentially reducing the time needed for environmental review, required under US environmental laws (Marcucci and Jordan 2013). Larger parcels are also more ecologically effective (Soule et al 2003, Gaston et al 2006), meaning a RAMP approach can be better both environmentally and economically. In support of these concepts, the US Federal Highways Administration endorses federal, state and local government applications of RAMP in transportation planning through a program called 'Eco-Logical[1]', (Brown 2006). In California, our study area's state, RAMP approaches are being used by a few individual county-level governments, and efforts to implement a state-level program are ongoing (Greer and Som 2010)[2]. A similar approach has been developed by The Nature Conservancy called 'Development by Design' that combines scientific assessment of infrastructure impacts with landscape-level conservation planning[3].

As a framework, RAMP engages multiple parties, including transportation infrastructure developers (governmental or private), government agencies that enforce environmental laws, and research and environmental groups that have identified critical lands for conservation. This cooperation between multiple stakeholders permits the incorporation of planning principles and environmental considerations early in the development of transportation infrastructure and other construction plans and projects (Thorne et al 2009a, 2009b).

The framework used in RAMP is similar to EIA, which is an internationally recognized and often required tool for environmental management, with global (Wandersforde-Smith 1980) and regional reviews (EU: Barker and Wood 1999, Middle East and North Africa: El-Fadl and El-Fadel 2004, Brazil: Glasson and Salvador 2000). The EIA process requires proposed projects to comply with sustainable development goals. Initially EIA focused on the impacts on the natural environment. Today, and because of suggestions to give EIA a more deterministic role in planning processes (Jay et al 2007), EIA also includes impacts on the human environment (social and economic)[7]. RAMP deals with ecological and environmental concerns, as in the earlier forms of EIA.

Claims that EIA as a process was unable to include the three factors of sustainability (environmental, economic and social), and that these should be addressed earlier in the planning phase have led to the development of strategic environmental assessments (SEA, Partidário 2000). The SEA can be thought of as the counterpart to the framework component of RAMP. In Europe, the SEA framework is intended to be used to ensure that impact assessment and sustainability are accounted for during planning phases. The SEA framework advocates early EIA, and suggests that it can (1) elevate environmental concerns to the same level as other aspects of development, (2) promote multi-organizational communication and collaboration, (3) be scaled from project to region, and (4) develop codes of conduct for mitigation and compensation. To do so, the SEA framework strengthens and streamlines the EIA process by the early identification of potential impacts and cumulative effects (Partidário 2000). The SEA process has been widely applied (Chaker et al 2006, Alshuwaikhat 2005, Briffet et al 2003); however, SEA procedures are not yet fully functional because they often do not progress beyond the planning phase to implementation (Bina

2007, Wallington et al 2007). In a SEA planning phase there is a need to do spatially cumulative assessments of multiple projects (cumulative effects assessment[8] —CEA, Gunn and Noble 2011). One possible way by which SEA could progress from strategy to implementation would be through the use of a RAMP framework. RAMP's project-based impact assessments are intended to both integrate environmental information into project planning and development to potentially reduce impacts on the environment from planned transportation projects and, as a basis for identifying what regional areas could be used to compensate for the estimated impacts during environmental mitigation. This process can be implemented early in, or prior to, the implementation of the transportation projects.

The RAMP framework also provides a way to integrate funds, when those are required of road developers (governmental or private) for environmental mitigation, with regional ecological and conservation goals. This coordination between transportation development and other regional objectives can be implemented for example, by pooling environmental mitigation funds from several transportation projects, to enable the purchase of larger parcels, which produces better results for ecosystem processes, viable species populations and their habitats (Soule et al 2003, Gaston et al 2006). The ecological effectiveness of the environmental mitigation can be further enhanced by coordinating mitigation actions, such as targeting the purchase of lands for preservation, within regional conservation priorities, here called a Greenprint (figure 1).

The RAMP methods use spatial assessment tools to determine regional environmental impacts from multiple transportation projects (left-hand column of figure 1), and to coordinate the locations of environmental mitigation for those projects with regional conservation objectives. The regional impact assessment can be compared and contrasted to impacts assessed on a project-by-project basis, to determine whether a regional approach will be ecologically and economically beneficial. RAMP includes a method to project the environmental mitigation needs for each individual and the aggregate impacts; for example, by including how much area of different land cover types may be needed, and where (within the region) mitigation sites can be found for acquisition. This step requires a map of the region that portrays land that has been targeted for conservation, restoration, ecosystem services or other objectives (right-hand column of figure 1). This integration of mitigation with conservation needs can provide an opportunity for better regional outcomes for environmental and social goals. This paper demonstrates the implementation of the first part of the RAMP approach for the transportation network in the Bay Area, and discusses the utility of the approach for environmental mitigation under different contexts, including whether a RAMP approach can help link SEA and EIA.

5.3 METHODS

The San Francisco Bay Area (hereafter called 'the Bay Area') region's nine-county population reached 7.15 million in 2005 from 1.73 million in 1940 (Department of Finance 2013), and is projected to grow to 10.22 million by 2050 (Public Policy Institute of California, PPIC 2009). In the Bay Area, regional government comprised of the Metropolitan Transportation Commission and the Association of Bay Area Governments (MTC and ABAG), is charged with developing a regional urban growth and transportation strategy. They published the 'Plan Bay Area' report, which identifies Bay Area transportation needs and a transportation infrastructure growth strategy over the next 25 years[6]. This report identifies 698 road and railroad transportation projects that are planned for construction in the 19 369 km² region (figure 2). We used the first phase of a RAMP framework (figure 3) to assess the potential aggregate impacts to protected species and their habitats, and regional concerns, by assessing impacts to agricultural lands (food security and economy) and wetlands (ecosystem process).

We reviewed the 698 planned transportation projects identified in GIS files provided by the MTC, using aerial imagery (Google Maps 2013[7]), and identified 181 road and railroad projects that crossed non-urbanized land or significant streams for inclusion in the regional impacts assessment (figure 4). Projects with predominantly urban impacts were excluded. We used ArcGIS v10 (ESRI, Redding CA) to assess aggregate environmental impacts. For each project, we identified impacts to species of concern and their habitats, to major natural vegetation types, and to agricultural lands.

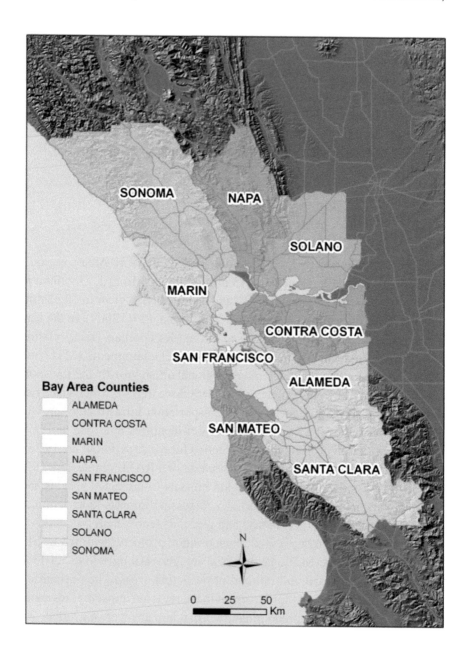

FIGURE 2. The nine-county San Francisco Bay region.

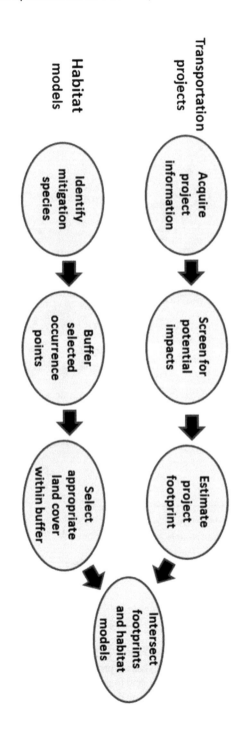

FIGURE 3: A flowchart of the environmental impact assessment methods used.

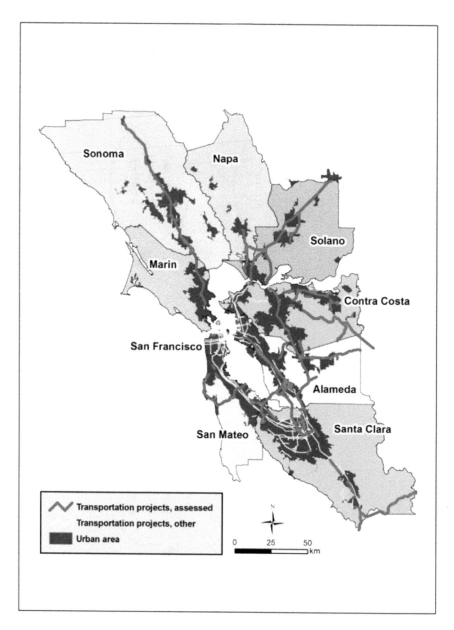

FIGURE 4: The San Francisco Bay Area planned transportation projects included in the analysis. Projects in red were analyzed to project impacts to threatened or endangered species, to landcover and agriculture, and to anadromous fish-bearing streams. Projects in yellow are in urban areas and were not analyzed.

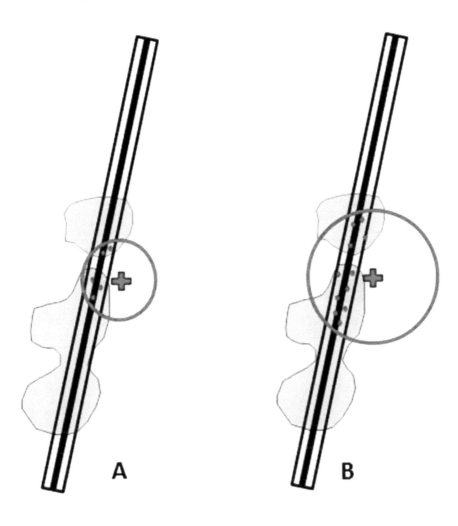

FIGURE 5: The use of species location records with project impact footprints, to identify the extent of suitable habitat that could be impacted. Green represents potential habitat for a species. The red cross represents a known occurrence record for that species. The black outline depicts the estimated project footprint. The red dots show the area within 1.6 and 3.2 km buffers of known locations that is within a project footprint and contains potential suitable habitat. These habitat extents were summed to provide the impact projection for that species on that project.

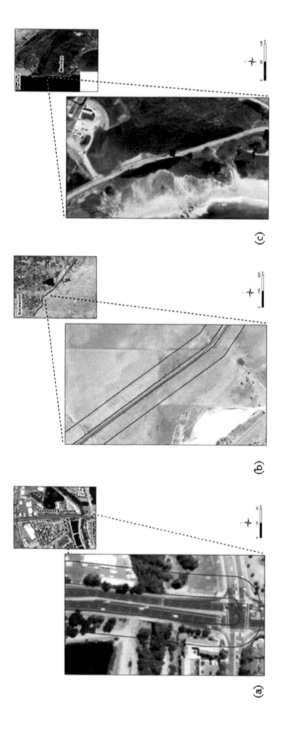

FIGURE 6: Three individual transportation project impact assessments. Each projects extent is shown in the upper image, and the area selected by inner and outer buffers is shown in the lower image. The area inside the inner buffer depicts the existing roadway, while the outer buffer depicts the estimated new construction impacts. Projects are (a) RTP22175, (b) RTP22400, and (c) RTP240114.

Calculation of transportation project footprints was accomplished in two ways. For new projects, the width and length of the new project was used to obtain a map of the area that may be impacted (figure 5). For modifications to existing transportation infrastructure (the majority of projects in our area), we used two buffers from the project's centerline or center point (figure 6). For the inner buffer, we used one meter square aerial imagery[8] to measure the typical distance from the centerline to the edge of the paved surface for each project. The centerline was then buffered by this distance. The second buffer is the distance from the edge of the inner buffer to the edge of the new project, which is the zone of new impact that different types of projects (e.g. adding highway lanes) will occupy when constructed (Thorne et al 2009a). The inner area was then buffered by the second distance to arrive at a total project footprint (e.g. figure 6(a)). For rail projects, we consulted an engineer from the Bay Area Rapid Transit (BART) program, to determine what the typical widths of impacts for different types of rail and terminal projects would be and applied those as buffers.

We then overlaid the transportation project footprints on a conservation greenprint to determine the suitable habitat that may be impacted within the project footprints (second row of figure 3). In order to develop the conservation greenprint, we compiled spatial data on the documented locations of threatened and endangered species, maps of their habitats, maps of natural landcover types and streams, and the locations of important agricultural lands. We obtained the recorded locations of threatened and endangered species that occurred within and up to eight kilometers beyond the study region from the state of California's Natural Diversity Data Base, CNDDB[9] and selected records after 1980. We used a habitat modeling approach by identifying the habitats used by each species, and developing maps of their suitable habitat from several different landcover maps. Habitats, also called vegetation types, used by California terrestrial vertebrate species are defined by a classification called the California Wildlife Habitat Relationships (CWHR), published by the California Department of Fish and Wildlife[10]. The appropriate habitats for each species were then selected from the landcover maps to portray the suitable locations for each species across our study area. For plant species, we used a similar process to identify suitable habitats, with a floristic database,

Calflora[11]. Various online sources for invertebrate species were used to define appropriate habitat types. Plant and invertebrate species' habitat requirements were then linked to the CWHR types represented in the maps of the region, so that their potential locations on the landscape could be identified using the reference maps. Using the following list of landcover maps, maps of suitable habitats for every terrestrial species were developed: the Conservation Lands Network[12], the National Wetlands Inventory[13], the San Francisco Estuary Institute's Inventory of wetlands[14], and the Great Valley Vernal Pool Distribution (Holland 2009).

Once the project footprints and habitat extents were established, we could measure the types of impacts and their extent. We buffered the known locations of each species by 3.2 and 6.4 (two and four miles), and selected the suitable habitat map area inside these areas (figure 5). These buffered habitat maps were then intersected with the footprints of the transportation projects to project the extent of habitat in each transportation project's footprint to be labeled as impacted-area habitats (figure 5). When the length of a transportation project extends beyond the area identified by the species buffer, there could be additional suitable habitat inside the transportation footprint, but beyond the species location buffer. Since we do not have actual evidence that the habitat is occupied, we limited the extent of suitable habitat selected to inside the species location buffers (the circles in figure 5). This makes our projections of impacts more conservative than by selecting all suitable habitat regardless of its distance from a point of known occurrence, and recognizes that not all suitable habitat for a species is occupied. Species habitat and agricultural location data were overlaid with project footprints to derive likely location and extent of impacts to species and habitats due to project construction (figure 5). Impacts were summed by species and aggregated to county and regional scales.

In addition, species-specific habitat maps for aquatic species were assembled. These included Delta smelt (*Hypomesus transpacificus*) and salmonid fish distribution data and critical habitat requirements from the US Fish and Wildlife Service[15]. We examined impacts to anadromous fish habitat by reviewing the number of times planned projects crossed waterways designated as critical habitat for these fish species. Typically, impacts to fish from transportation projects occur not from habitat loss but from direct mortality to individuals, especially juveniles, during the construction period itself. While temporary habitat loss on site is generally restored, additional offsite actions to mitigate for loss of individuals is usually required. Often, these actions include removal of in-channel barriers to fish movement in the vicinity of the impacts. If there are multiple impacts within a watershed, mitigation requirements and funding could conceivably be bundled for several projects, potentially leading to a greater ecological return on the mitigation investment. While the identification of aquatic mitigation locations is beyond the scope of this paper, a critical first step is the assessment of the impact locations, which we have included.

Finally, impacts to agricultural lands[16] and to some natural vegetation types (oak woodlands, riparian forest, and wetlands) were identified because these landcover types require mitigation regardless of the presence of listed species. In addition, loss of agricultural land is potentially a food security issue and reduces regional agricultural economic output; and loss of water habitats has consequences for ecosystem processes and biodiversity. For these, we calculated the full extent of impacts within the boundaries of each transportation project.

5.3.1 REGIONAL VERSUS INDIVIDUAL PROJECT IMPACT ASSESSMENT

Typically, environmental impacts are assessed on a project-by-project basis, and subsequent mitigation to offset any impacts is also conducted on this basis. We present impact assessments from three individual projects for comparison with the regional assessment. The impacts were assessed using the same approach as for the regional assessment, and the results are also part of the summed regional results. The first project is RTP 22175, located in Santa Clara county titled, "Widen Almaden Expressway from Coleman Avenue to Blossom Hill Road" (figure 6(a)). This project is 1.1 km in length. The assigned buffers were 12 m (existing road) and 50 m (proposed project). The second project, RTP 22400 is located in Contra Costa county titled, "Conduct environmental and design studies to create a new alignment for SR239 and develop corridor improvements from Brentwood to Tracy—project development" (figure 6(b)). This project is 17.5 km in length, among the larger projects assessed overall. The assigned buffers were 20 m (existing road) and 152 m (proposed project). The third project is RTP 240114, located in San Mateo county titled, 'Implement operational and safety improvements on route 1 between half Moon Bay and Pacifica (includes acceleration lanes, deceleration lanes, turn lanes, bike lanes, and enhanced crossings)' (figure 6(c)). This project is 28.8 km in length. The assigned buffers were 7 m (existing road) and 15 m (proposed project).

5.4 RESULTS

We analyzed 181 transportation projects that will cross natural lands and streams (table 1, figure 4), and found that 107 (59%) will have impacts to listed species, ecosystems, or agriculture. Of these, eight are rail and the others road projects. The aggregate rail projects' footprint area is 1551 ha as compared with 3372 ha for roads.

TABLE 1: A list of the number of projects in each county and regionally. Note is a project crosses county lines, it is counted in each county where it occurs.

Region or type of project	Number of projects listed in MTC report	Number of projects included in analysis	Projected total costs for selected projects (in millions)
Bay Area region/ multi-county	72	20	$5882
Alameda county	117	36	$4752
Contra Costa county	106	39	$2158
Marin county	29	a	—
Napa county	20	15	$299
San Francisco county	54	a	—
San Mateo county	58	15	$733
Santa Clara county	155	33	$10 989
Solano county	46	10	$989
Sonoma county	31	13	$614

ᵃ*For these counties only regional projects were included in the analysis.*

TABLE 2: Area impacts by species habitat from transportation projects (hectares).

Area impacts by species habitat		Impacts: 3.2 km (in hectares)	Impacts: 6.4 km	No. projects: 3.2 km	No. projects: 6.4 km
Acanthomintha duttonii	San Mateo thorn-mint	1.66	2.75	2	3
Agelaius tricolor	Tricolored blackbird	278.2	552.36	23	32
Ambystoma californiense	California tiger salamander	474.21	509.38	38	44
Amsinckia gran-diflora	Large-flowered fiddleneck	8.54	19.8	1	3
Athene cunicu-laria	Burrowing owl	344.4	493.5	41	53
Blennosperma bakeri	Sonoma sunshine	7.7	7.7	1	1
Branchinecta conservatio	Conservancy fairy shrimp	0.73	13.4	1	1
Branchinecta longiantenna	Longhorn fairy shrimp	0.0	1.5	—	3
Branchinecta lynchi	Vernal pool fairy shrimp	41.8	41.8	6	6
Buteo swainsoni	Swainson's hawk	324.4	559.5	20	23
Callophrys mos-sii bayensis	San Bruno elfin butterfly	6.1	7.6	3	3
Castilleja affinis ssp. neglecta	Tiburon paintbrush	1.6	25.2	2	9
Ceanothus fer-risiae	Coyote ceanothus	0.0	0.04	—	1
Chloropyron molle ssp. molle	Soft bird's-beak	0.0	<0.01	—	1
Chloropyron palmatum	Palmate-bracted bird's-beak	2.2	7.9	3	7
Cirsium fontin-ale var. fontinale	Fountain thistle	0.0	<0.01	—	1
Dudleya abram-sii ssp. setchellii	Santa Clara valley dudleya	0.04	0.04	1	1
Elaphrus viridis	Delta green ground beetle	0.0	6.8	—	1
Eriophyllum latilobum	San Mateo woolly sunflower	0.0	0.45	1	2
Eryngium rac-emosum	Delta button-celery	0.0	3.36	—	1

TABLE 2: *Cont.*

Area impacts by species habitat		Impacts: 3.2 km (in hectares)	Impacts: 6.4 km	No. projects: 3.2 km	No. projects: 6.4 km
Hesperolinon congestum	Marin western flax	0.08	0.08	1	1
Lasthenia burkei	Burke's goldfields	0.0	0.04	—	1
Lasthenia conjugens	Contra Costa goldfields	21.5	21.5	2	2
Laterallus jamaicensis coturniculus	California black rail	43.1	45.9	12	21
Lepidurus packardi	Vernal pool tadpole shrimp	15.4	21.5	2	2
Limnanthes vinculans	Sebastopol meadowfoam	0.0	0.12	—	2
Masticophis lateralis euryxanthus	Alameda whipsnake	0.0	0.32	—	1
Navarretia leucocephala ssp. plieantha	Many-flowered navarretia	5.9	7.4	1	1
Plebejus icarioides missionensis	Mission blue butterfly	5.5	6.4	2	3
Pleuropogon hooverianus	North Coast semaphore grass	0.0	0.04	—	1
Potentilla hickmanii	Hickman's cinquefoil	0.16	0.24	1	1
Rallus longirostris obsoletus	California clapper rail	121.3	215.3	19	25
Rana draytonii	California red-legged frog	250.9	345.9	42	51
Reithrodontomys raviventris	Salt-marsh harvest mouse	10.9	19.7	7	7
Riparia riparia	Bank swallow	0.0	0.16	—	1
Speyeria callippe callippe	Callippe silverspot butterfly	0.0	5.6	—	2
Sternula antillarum browni	California least tern	0.24	6.5	1	5
Streptanthus albidus ssp. albidus	Metcalf Canyon jewel-flower	0.04	0.04	1	1

TABLE 2: *Cont.*

Area impacts by species habitat		Impacts: 3.2 km (in hectares)	Impacts: 6.4 km	No. projects: 3.2 km	No. projects: 6.4 km
Suaeda califor-nica	California seablite	5.02	5.3	1	1
Thamnophis gigas	Giant garter snake	0.0	0.24	—	1
Thamnophis sir-talis tetrataenia	San Francisco garter snake	3.68	7.28	3	6
Trifolium amoe-num	Showy rancheria clover	0.65	10.08	4	6
Vulpes macrotis mutica	San Joaquin kit fox	249.9	332.41	15	25
Total species impacts	—	2411.4	3490.2	107	107

The estimated impacts from the 107 projects include 43 terrestrial threatened and/or endangered species were recorded within 6.4 km of the project extents, and 30 within 3.2 km (table 2), including 20 plant species, seven birds, five amphibians and reptiles, four insects, two mammals, and four aquatic invertebrates. Impacted species were affected by an average of 8.4 projects using the 6.4 km buffer and 6.4 projects using the 3.2 km buffer (table 2).

Seventy seven projects were projected to have 990 ha of direct land-cover impacts to farmland, 17 projects impacted 13 ha of riparian vegetation, 19 projects impacted 37 ha of oak woodlands, 36 projects impacted 70 ha of wetlands, eight projects impacted 50 ha of vernal pools, and 16 projects impacted 16 ha of open water, totaling 1175 ha of land-cover impacts.

Fifty eight projects contained 125 stream crossings that may affect four categories of threatened or endangered salmonid fish: Chinook salmon (*Oncorhynchus tshawytscha*), and steelhead (*Oncorhynchus mykiss*), which come from three evolutionarily significant units, the Central Coast, Central Valley and South-central coast, representing uniquely composed genetic subpopulations (Busby et al 1996).

TABLE 3: The projected area of suitable habitat impacted (ha) within the buffered footprints for the three example projects, listed by their project identification number.

Species	Project 22175 (ha)	Project 22400 (ha)	Project 240114 (ha)
California red-legged frog	0.0–4.17	71.99–73.69	15.5
Tricolored blackbird	—	146.8–312.9	—
California tiger salamander	—	164.38	—
Burrowing owl	—	119.9–163.3	—
Vernal pool fairy shrimp	—	18.6	—
Swainson's hawk	—	208.78–350.58	—
Delta button-celery	—	0.0–3.4	—
San Joaquin kit fox	—	163.3	—
San Bruno elfin butterfly	—	—	5.63–6.07
Mission blue butterfly	—	—	0.0–0.85
Hickman's cinquefoil	—	—	0.16–0.24
San Francisco garter snake	—	—	2.75–3.76
Open water	<0.04	0.16	—
Wetlands	0.12	3.92	0.24
Oaks	—	—	0.12
Riparian forest	—	—	0.08
Farmland	—	278.3	5.58
Vernal pools	—	1.86	—

5.4.1 REGIONAL VERSUS INDIVIDUAL PROJECT IMPACT ASSESSMENT

Of the 43 threatened or endangered species, 58% are projected to be impacted by more than one project. Taking examples by taxonomic group and using the smaller (more conservative) buffer analysis, we find that: *Rana draytonii* (California red-legged frog) habitat is found in 42 projects, with habitat impacts ranging from 0.02–72 ha; *Athene cunicularia* (burrowing owl) habitat is found in 41 projects with habitat impacts ranging from 0.04–119.3 ha per project; *Vulpes macrotis mutica* (San Joaquin kit fox) habitat is found in 15 projects with habitat impacts of 0.03–163 ha per project; *Branchinecta lynchi* (vernal pool fairy shrimp, that lives in seasonal wetlands) habitat is found in six projects with habitat impacts ranging from 0.16–20.9 ha per project; and, *Castilleja affinis* ssp. *neglecta* (Tiburon paintbrush) is found in two projects for >0.01 and 1.6 ha extent of habitat impacts.

To compare regional impacts to those by project, we used the three example projects. Twelve threatened or endangered species are expected to have habitat impacted from at least one of the three individual projects (table 3). Some very limited-distribution species, such as *Potentilla hickmanii* or *B. lynchi,* only appear in the analysis for these three projects. However 83% of impacted species from the three projects are also impacted by other projects. The large single example project, TRP 22400, impacts eight species, one uniquely, and contributes between 35–65.3% of the habitat impacted for the other seven to the regional total. Habitat for these seven species is also impacted by from 1–53 other projects. Similarly, all of the major ecosystems impacted by the three projects (table 3) are also impacted by between five and 74 other planned projects.

5.5 DISCUSSION

5.5.1 REGIONAL AND PROJECT LEVEL ASSESSMENTS

Nearly 24% (1175 ha) of the planned transportation projects analyzed in the Bay Area are projected to impact landcover types that require mitigation in California. Additionally, there are spatially cumulative impacts to habitats that support between 30 and 43 threatened and endangered species, totaling between 2411–3480 ha. Over half of these species will be impacted by more than one project. The environmental mitigation for these spatially cumulative, regional impacts will likely require considerable open space (>30 km²). In contrast, the project-by-project analysis indicated only minor impacts to species and habitats (three to four orders of magnitude smaller). For example, burrowing owls (*A. cunicularia*) are expected to have 344–494 ha of habitat impacted by 41–53 projects. Project-by-project mitigation for this species would yield a myriad of small, potentially poorly performing habitat reserves. Combining the mitigation solutions from the projects impacting this species could greatly reduce the number of individual land acquisition transactions required (with associated cost and time savings), and yield larger, potentially better functioning reserves. Additionally, the aggregate loss of nearly 10 km² of farmland to 77 projects, and 70 ha of wetlands to 36 projects shows how expansion consumes landcover types which provide important ecosystem services such as food and water retention. By considering the spatially aggregate impacts to landcover types, it may be possible to put larger blocks of remaining lands under conservation management which would protect them from conversion, and assist in provisioning the region with an important base of ecosystem services. These results show there are compelling reasons to consider the impacts from a spatially cumulative perspective, which offer the possibility to conduct better environmental practice. Government regulatory representatives will find environmental mitigation better meeting the needs of species, and transportation agencies will potentially find economic savings through the purchase of fewer, larger properties, as well as recognition for improvement in environmental practice.

The use of a RAMP framework for providing compensatory mitigation for transportation projects appears justified because of the large number of projects and the extent of the expected impacts. Further, the mitigation required by government environmental regulation agencies to permit the development of these projects will likely be more than just the area of direct impacts we quantified. Regulatory agencies often require multipliers for impacts to calculate mitigation needs. These multipliers vary by the type of impact and the ecological context, with (in the US) no net loss laws requiring a minimum of 1 ha of wetlands constructed for every 1 ha developed, and typically additional extant wetlands protected (US Clean Water Act 1972[17]). In California, many habitat types for endangered species may require a higher ratio. For example, blue oak woodlands (*Quercus douglassii*) are frequently assessed at a 3:1 ratio, whereby three hectares must be purchased or restored for each hectare impacted.

The development of regional EIAs, and subsequent identification of appropriate locations to accommodate the environmental offsets, is at the heart of California's RAMP framework. An interesting phenomenon is that groups using the approach are emerging at several spatial scales and administrative levels, such as federal initiatives advanced through Ecological (Brown 2006) and two county-level initiatives in California. One measure of success of the approach is where conservation lands have actually been acquired to offset transportation infrastructure impacts, in other words by the completion of the RAMP steps. The two county-level initiatives in California are doing this to offset spatially cumulative impacts from multiple transportation projects well in advance of transportation project delivery. San Diego county has a program that uses funds from a county sales tax measure to pay for offsets from transportation project impacts[18], while Orange county (near Los Angeles) has acquired at least four properties to address similar impacts, also funded through a county sales tax measure[19]. This suggests the utility of the approach, and that it would be possible to implement it elsewhere, even while other areas have different environmental laws, administrative systems and capacities.

5.5.2 RAMP AND SEA

The implementation of RAMP in some counties in California, the results from the Bay Area impact assessment, and the national guidelines promoting a RAMP framework may be informative for European groups that are attempting to implement SEAs. The challenge for the EU has been how to implement spatially cumulative effect assessments (CEAs) which are the phase of a SEA where implementation occurs. The RAMP framework shows one way that this integration of planning, environmental impact analysis, project construction and environmental mitigation can be operationalized. While the specifics of environmental laws, business practice, and environmental practice differ between the EU and the US, the underlying needs and challenges are the same.

This study illustrates how RAMP could act as the operationalization tool to bring SEA from a strategic to an implementation level. We show that RAMP addresses spatially cumulative impacts (for example cumulative impacts on California red-legged frogs and burrowing owl, for which project-specific impacts are much smaller than those of the RAMP spatially cumulative estimated impacts). We also demonstrate that RAMP is more than that. It can be the step at a regional level that is necessary to bridge between the planning level of SEA and the project-by-project level of an EIA. In the Bay Area assessment, we show that 58% of the threatened and endangered species are projected to be impacted by more than one project. Thus, the scaling from individual to multiple project assessments of impact for those species allows for a regional planning of mitigation and compensation measures, likely more effective than project-by-project planning from both a transportation delivery and ecological perspective. RAMP can thus be coupled with the assessment of mitigation and compensation needs (how much is required from the sum of all projects); it can be used in partnership with systematic conservation planning tools for conservation land identification (where best can we meet those requirements); and because of this regional perspective, RAMP can identify the best strategy for conservation land acquisition and designation (how to best use mitigation funds to acquire land for conservation).

5.5.3 RAMP IN OTHER METROPOLITAN REGIONS

The population growth projections in the Bay Area (PPIC 2009), and for cities globally, (McDonald et al 2011) suggest that cities will continue to expand. Further, the urban extent in the Bay Area has expanded by over 1842 km^2 since 1940, as population grew from 1.7 to 7.1 million (Thorne et al 2013), trends that are similar to hundreds of other cities globally, where populations have on average doubled and area used has tripled (Schneider and Mertes 2014) or quadrupled over the past 20 years (Deto et al 2011). The pressures these trends and associated transportation infrastructure will put on natural ecosystems suggest the need is urgent, and that a RAMP framework may be an attractive option for planners in other metropolitan regions who seek better outcomes for conservation and infrastructure development. Infrastructure agencies are a promising group for environmental groups to work with to initiate this type of planning because of their long planning horizons (Thorne et al 2009b), and it is hoped that by setting an example, these agencies may also be able to influence environmental mitigation practices from other types of urban growth impacts, particularly housing construction. The Bay Area implementation of the RAMP framework will need to involve multiple jurisdictions, which may be typical of what would be encountered in other metropolitan regions.

In many developing countries there are fewer legal obligations for environmental offsets (Kuokkanan 2002), and in many cases, less digital data are available about the locations of the resources and species of concern. In such areas, a RAMP approach could potentially still be useful as a way to develop regional planning. First, its use would require an evaluation of the data that are lacking to conduct an EIA of a road network, which could provide government and environmental NGOs with an agenda about what type of information is needed to be developed. Second, even without specific information on the locations of species, impacts to landscape connectivity, to linear features such as rivers, and to the general landcover types, as identified by coarse-resolution landcover maps (for example, the CORINE program from the European Environmental Agency, or MODIS products), can be assessed. The RAMP approach can even be applied in regions that have very little data or capacity, through recognition that the development of road networks has spatially cumulative impacts. This recognition can provide the basis to develop partnerships between those responsible for the implementation of roads, and those with the knowledge and expertise to assess what impacts might arise.

5.5.4 RAMP SHORTCOMINGS/LIMITATIONS

Barriers to the application of a complete RAMP framework include data and knowledge, skills and capacity building, institutional configurations and legislative power. Like SEA (João 2007), RAMP can be a data hungry platform and the spatial extent of RAMP projects needs to be defined, to understand the issues that are at stake. This data need is exacerbated by the assessment of cumulative effects (Therivel and Ross 2007) and the across-scales interactions emerging from this regional approach (Partidário 2007). Local or federal governments, universities, and NGOs are the primary repositories of such data and knowledge, and need to be engaged in the process. Some federal US agencies are interested in putting these data online to promote this type of planning, for example, the US Federal Highways ESA WebTool[20], and the Environmental Protection Agency's NEPAssist tool[21]. However, such tools are relatively new, are not populated with data or the data are incomplete depending on the state, or are not yet widely available.

The RAMP framework requires some investment in planning and design that is typically earlier in the cycle of project development than is the case under business as usual. Therefore, a key consideration for convincing regional governments to adopt the approach is to demonstrate that there are enough biological and environmental impacts from new infrastructure projects to justify implementing a regional environmental mitigation, by identifying suitable environmental offsets to satisfy obligations from multiple projects. This can be hindered by institutional configurations and legislative power. Institutions have agendas of their own and clear jurisdictional boundaries necessary to track accountability in decisions. Cross-institutional collaboration is rare, but RAMP provides a framework for involvement of multiple stakeholders in the planning process, wherein each group has something to gain through the collaboration. The challenge ahead is how to promote institutional configurations that allow for collaboration while preserving accountability. Different agencies comply differently to legislation and regulation at different governance levels (for example international conventions like the Ramsar convention, US national laws like the Endangered Species Act, or local policies like the California Lands Act). The lack of environmental regulation need not to be a barrier to RAMP, as a code of best conduct can be implemented, or NGOs can play a significant role in managing environmental needs.

5.6 CONCLUSION

This analysis demonstrated the screening of 698 transportation projects in a major metropolitan area to identify 181 that cross open space, and found that 107 of those will have cumulative biological and environmental impacts of 3586–4655 ha. These impacts will affect at least 30 threatened or endangered species, over half of which will be impacted by more than one project. These results point to the need to address the environmental mitigation for these impacts in aggregate. The RAMP framework offers an opportunity to do so, with potential for better ecological results, and that the implementation may be less expensive than project-by-project mitigation. In the Bay Area, assembling a round table of transportation agency and environmental regulation agency representatives will permit discussion of steps for implementation of a RAMP. In other parts of the world, the RAMP framework offers a model for development of cooperative regional planning that can result in better environmental outcomes.

FOOTNOTES

1. http://www.environment.fhwa.dot.gov/ecological/eco_index.asp.
2. https://rampcalifornia.water.ca.gov/web/guest.
3. http://www.nature.org/ourinitiatives/urgentissues/smart-development/publications/index.htm.
4. http://www.unep.org/regionalseas/publications/reports/RSRS/pdfs/rsrs122.pdf.
5. http://www.imperial.ac.uk/pls/portallive/docs/1/21559696.PDF.
6. http://onebayarea.org/pdf/Draft_Plan_Bay_Area_3-22-13.pdf.
7. https://maps.google.com/maps.
8. http://www.fsa.usda.gov/Internet/FSA_File/naip_2009_info_final.pdf.
9. http://www.dfg.ca.gov/biogeodata/cnddb/.
10. http://www.dfg.ca.gov/biogeodata/cwhr/wildlife_habitats.asp.
11. http://www.calflora.org/.
12. cln_veg; http://www.bayarealands.org/.
13. http://www.fws.gov/wetlands/.
14. http://www.sfei.org/ecoatlas/gis.
15. http://ecos.fws.gov/crithab/.
16. http://conservation.ca.gov/dlrp/fmmp/Pages/Index.aspx.
17. Title 33 U.S.C. Part 1344. http://www.epw.senate.gov/water.pdf.
18. http://www.sandag.org/uploads/publicationid/publicationid_1138_4880.pdf.
19. http://www.octa.net/Measure-M/Environmental/Freeway-Mitigation/Overview/.
20. http://www.environment.fhwa.dot.gov/esawebtool/.
21. http://nepassisttool.epa.gov/nepassist/entry.aspx.

REFERENCES

1. Alshuwaikhat H M 2005 Strategic environmental assessment can help solve environmental impact assessment failures in developing countries Environ. Impact Assess. Rev. 25 307–17
2. Barker A and Wood C 1999 An evaluation of the Environmental Impact Assessment system performance in eight EU countries Environ. Impact Assess. Rev. 19 387–404
3. Barthelemy M, Bordin P, Berestycki H and Gribaudi M 2013 Self-organization versus top-down planning in the evolution of a city Sci. Rep. 3 2153
4. Bina O 2007 A critical review of the dominant lines of argumentation on the need for strategic environmental assessment Environ. Impact Assess. Rev. 27 585–606

5. Briffett C, Obbard J P and Mackee J 2003 Towards Strategic Environmental Assessment for the developing nations of Asia Environ. Impact Assess. Rev. 23 171–96
6. Brown J W 2006 Eco-Logical: An Ecosystem Approach to Developing Infrastructure Projects (Washington, DC: Office of Project Development and Environmental Review, Federal Highway Administration) www.environment.fhwa.dot.gov/eco-logical/eco_index.asp
7. Busby P J, Wainwright T C, Bryant G J, Lierheimer L J, Waples R S, Waknitz F W and Lagomarsino I V 1996 Status review of west coast steelhead from Washington, Idaho, Oregon, and California NOAA Technical Memorandum NMFS-NWFSC-27 www.westcoast.fisheries.noaa.gov/publications/status_reviews/salmon_steelhead/steelhead/sr1997-steelhead0.pdf
8. Chaker A, El-Fadl K, Chamas L and Hatjian B 2006 A review of strategic environmental assessment in 12 selected countries Environ. Impact Assess. Rev. 26 15–56
9. Department of Finance 2013 Historical US Census Populations of Counties and Incorporated Cities/Towns in California: 1850–2010 (Sacramento, CA: California State Data Center)
10. Deto K C, Günerlap F M and Reilly M K 2011 A meta-analysis of global urban land expansion PLoS ONE 6 e2377
11. El-Fadl K and El-Fadel M 2004 Comparative assessment of Environmental Impact Assessment systems in MENA countries: challenges and prospects Environ. Impact Assess. Rev. 24 553–93
12. Epps C W, Palsboll P J, Wehausen J D, Roderick G K, Rmaey R R II and Mc-Cullough D R 2005 Highways block gene flow and cause a rapid decline in genetic diversity of desert bighorn sheep Ecol. Lett. 8 1029–38
13. Forman R T T and Alexander L E 1998 Roads and their major ecological effects Ann. Rev. Ecol. Syst. 29 207–31
14. Forman RTT et al 2003 Road Ecology: Science and Solutions (Washington: Island Press)
15. Fuglestvedt J, Berntsen T, Myhre G, Rypdal K and Skeie R B 2008 Climate forcing from the transport sectors Proc. Natl Acad. Sci. 105 454–8
16. Gaston K J et al 2006 The Ecological Effectiveness of Protected Areas: The United Kingdom 132 76–87 Biol. Conservation
17. Gelbard J L and Belnap J 2003 Roads as conduits for exotic plant invasions in a semiarid landscape Conserv. Biol. 17 420–32
18. Girvetz E H, Thorne J H, Berry A M and Jaeger JAG 2008 Integration of landscape fragmentation analysis into regional planning: a statewide multi-scale case study for California Landscape Urban Plan. 86 205–18
19. Glasson J and Salvador N N B 2000 Environmental Impact Assessment in Brazil: a procedures-practice gap. A comparative study with reference to the European Union, and especially the UK Environ. Impact Assess. Rev. 20 191–225
20. Gordon A, Bastin L, Langford W T, Lechner A M and Bekessy S A 2013 Simulating the value of collaboration in multi-actor conservation planning Ecol. Model. 249 19–25
21. Greer K and Som M 2010 Breaking the environmental gridlock: advance mitigation programs for ecological impacts Environ. Pract. 12 227–36

22. Grimm N B et al 2008a The changing landscape: ecosystem responses to urbanization and pollution across climatic and societal gradients Front. Ecol. Environ. 6 264–72

23. Grimm N B et al 2008b Global change and the ecology of cities Science 319 756–60

24. Gunn J and Noble B F 2011 Conceptual and methodological challenges to integrating Strategic Environmental Assessment and cumulative effects assessment Environ. Impact Assess. Rev. 31 154–60

25. Holland R F 2009 California's great valley vernal pool habitat status and loss: rephotorevised 2005 (Auburn, CA: Placer Land Trust)

26. Huber P R, Greco S E and Thorne J H 2010a Spatial scale effects on conservation network design: trade-offs and omissions in regional versus local scale planning Landscape Ecol. 25 683–95

27. Jaeger J et al 2005 Predicting when animal populations are at risk from roads: an interactive model of road avoidance behavior Ecol. Model. 185 329–48

28. Jay S, Jones C, Slinn P and Wood C 2007 Environmental impact assessment: retrospect and prospect Environ. Impact Assess. Rev. 27 287–300

29. Joao E 2007 A research agenda for data and scale issues in Strategic Environmental Assessment (SEA) Environ. Impact Assess. Rev. 27 479–91

30. Kark S, Levin N, Grantham H S and Possingham H P 2009 Between-country collaboration and consideration of costs increase conservation planning efficiency in the mediterranean basin PNAS 106 15368–73

31. Kennedy C et al 2009 Greenhouse gas emissions from global cities Environ. Sci. Technol. 43 7297–302

32. Kuokkanen T 2002 International Law and Environment: Variations on a Theme (Norwell, MA: The Erik Castrén Institute of International Law and Human Rights, Kluwer Law International)

33. Marcucci D J and Jordan L M 2013 Benefits and challenges of linking green infrastructure and highway planning in the United States Environ. Manage. 51 182–97

34. McDonald R I et al 2011 Urban growth, climate change, and freshwater availability 108 6312–7

35. McKinney M 2008 Effects of urbanization on species richness: a review of plants and animals Urban Ecosystems 11 161–76

36. Moilanen A, Anderson B J, Arponen A, Pouzols F D and Thomas C D 2013 Edge artifacts and lost performance in national versus continental conservation priority areas Divers. Distributions 19 171–83

37. National Research Council 2005 Assessing and Managing the Ecological Impacts of Paved Roads (Washington, DC: National Academies Press)

38. OCTA 2006 Orange County Transportation Authority Ordinance No. 3 http://www.octa.net/uploadedFiles/Measure_M_2011/Overview/Resource_Library_Listings/Other_Resources/m2ordinance.pdf

39. Partidário M R 2000 Elements of an SEA framework—improving the added-value of SEA Environ. Impact Assess. Rev. 20 647–63

40. Partidário M R 2007 Scales and associated data - what is enough for SEA needs? Environ. Impact Assess. Rev. 27 460–78

41. PPIC 2009 Long-run Socioeconomic and Demographic Scenarios for California (Sacramento, CA: California Energy Commission)

42. SANDAG (San Diego Association of Governments) 2012 TransNet Environmental Mitigation Program Fact Sheet http://www.sandag.org/uploads/publicationid/publicationid_1138_4880.pdf

43. Satterthwaite D, McGranahan G and Tacoli C 2010 Urbanization and its implications for food and farming Phil. Trans. R. Soc. B 365 2809–20

44. Schwartz M W, Thorne J H and Viers J H 2006 Biotic homogenization of the California flora in urban and urbanizing regions Biol. Conserv. 127 282–91

45. Seo C, Thorne J H, Choi T, Kwon H and Park C 2013 Disentangling roadkill: the influence of landscape and season on cumulative vertebrate mortality in South Korea Landscape Ecol. Eng.

46. Schneider A and Mertes CM 2014 Expansion and growth in chinese cities, 1978, 2010 Environ. Res. Lett. 9 024008

47. IOPscience

48. Soule M E, Estes J A, Berger J and Rio C M 2003 Ecological effectiveness: conservation goals for interactive species Conserv. Biol. 17 1238–50

49. Theobald D et al 2000 Incorporating biological information in local land-use decision making: designing a system for conservation planning Landscape Ecol. 15 35–45

50. Therivel R and Ross B 2007 Cumulative effects assessment: does scale matter? Environ. Impact Assess. Rev. 27 365–85

51. Thorne J, McCoy M, Hollander A, Roth N and Quinn J 2005 Regional Analysis for Transportation Corridor Planning Proc. of the Int. Conf. on Environment and Transportation (San Diego, CA: International Conference on Environment and Transportations) www.icoet.net/links.asp

52. Thorne J H, Gao S, Hollander A H, Kennedy J A, McCoy M, Johnston R A and Quinn J F 2006 Modeling potential species richness and urban buildout to identify mitigation sites along a California highway J. Transp. Res. D 11 233–314

53. Thorne J H, Girvetz E H and McCoy M C 2009a Evaluating aggregate terrestrial impacts of road construction projects for advanced regional mitigation Environ. Manage. 43 936–48

54. Thorne J H, Huber P R, Girvetz E H, Quinn J and McCoy M C 2009b Integration of regional mitigation assessment and conservation planning Ecol. Soc. 14 1–47

55. Thorne J H, Santos M J and Bjorkman J H 2013 Regional assessment of urban impacts on landcover and open space finds a smart urban growth policy performs little better than business as usual PLoS ONE 8 e65258

56. Trombulak S C and Frissell C A 2000 Review of ecological effects of roads on terrestrial and aquatic communities Conserv. Biol. 14 18–30

57. UNPF (United Nations Population Fund) 2007 UNPFA State of the World Population 2007 (New York, NY: United Nations) www.unfpa.org/webdav/site/global/shared/documents/publications/2007/695_filename_sowp2007_eng.pdf

58. Wallington T, Bina O and Thissen 2007 Theorising strategic environmental assessment: fresh perspectives and future challenges Environ. Impact Assess. Rev. 27 569–84

59. Wandersforde-Smith 1980 Environmental Impact Statements. a test model presentation UNEP Bangkok

CHAPTER 6

Quantifying the Total Cost of Infrastructure to Enable Environmentally Preferable Decisions: The Case of Urban Roadway Design

CONRAD A. GOSSE AND ANDRES F. CLARENS

6.1 INTRODUCTION

Ground transportation is responsible for nearly 30% of the primary energy consumption and 27% of the greenhouse gas (GHG) emissions in the United States [1]. Related infrastructure also results in significant material movement—every $1 million investment in roadway construction requires 9×10^4 tonnes of aggregate and 3.3×10^3 tonnes of cement [2]. In an effort to try to reduce these burdens, numerous life cycle assessments (LCAs) have been performed over the past decade to understand how specific technological choices contribute to emissions, energy consumption,

Quantifying the Total Cost of Infrastructure to Enable Environmentally Preferable Decisions: The Case of Urban Roadway Design. © *Gosse CA and Clarens AF.* Environmental Research Letters *8,1 (2013), http://dx.doi.org/10.1088/1748-9326/8/1/015028. Licensed under Creative Commons Attribution 3.0 Unported License, http://creativecommons.org/licenses/by/3.0/.*

and materials use. These studies have tended to focus on either the design or the use phase of the road [3]. Design generally involves the selection of a material, e.g., concrete or asphalt, or specification of roadway width and configuration [4]. Use entails a variety of other processes including vehicle selection or roadway maintenance [5]. Even though these studies have identified many obvious opportunities for environmental improvement, there is little evidence to suggest that they have provided deep reductions in material use or emissions.

A principal limitation of many published analyses is that they consider technological options for ground transportation narrowly and evaluate only specific elements of the design or use of roadways at once. An asphalt road may have a lower life cycle emission profile than a concrete road, for example, but that difference is small when compared to the overall emissions from the use phase of the road [6]. Similarly, the conclusion that a greater bicycle mode share will reduce the carbon emissions of a roadway is not useful if it is not considered along with other factors discouraging bicycle use and the impact of more bicycles on overall traffic flow. To date, LCA has been employed as a method for environmental bean counting that considers problems removed from the broader system within which they exist. Consequently, even though LCA has been actively pursued in academic circles, it has had only limited impact in policy circles.

At the same time that conventional LCA tools have been insufficient for solving many of the existing problems faced by transportation managers, emergent challenges make the need for new tools even more pressing [7]. Conventional development patterns have led to widespread congestion in urban and suburban areas around the world. Shrinking maintenance budgets at a time when facilities built in the post WWII boom period are reaching their design life span are making it ever harder to maintain the level of service that was envisioned for roadways during design. Declining pavement quality is also exacerbating the emissions [8] and safety costs associated with the use of these aging facilities, which only compound the impacts of unchecked growth in vehicle kilometers of travel (VKT) worldwide. Efforts to address these problems and provide meaningful improvements will require systems thinking that considers life cycle impacts, personal choice, and policy realities.

In practice, ground transportation is constrained by a few overarching factors. The most obvious is budget. Road construction is expensive, though less so than alternatives like public transport, because much of the cost is borne by users in the form of vehicles and fuel. Roadways are expensive to maintain, and so many exhibit condition ratings below their design values. In many urban areas, space is also constrained and a limiting factor in terms of enabling more mobility. Where space is available, additional road and parking facilities relieve congestion in the near term but only further separate typical destinations, increasing trip lengths over the long term [9]. The carbon emissions from ground transportation are appreciable and growing as more and more developed nations move toward car ownership levels on par with the United States. Paradoxically, the convenience afforded by automobiles has contributed to significant and consistent traffic-related delays in almost all of the worlds' urban centers. These delays translate into appreciable costs to the users of the transportation systems [10], in addition to increased crash and health risks [11], and overall environmental impacts.

Planning that tackles these challenges involves both near-term adaptive strategies and long-term improvement projects. Existing facilities, including the functionally obsolete, have significant embodied emissions and sunk costs that preclude their immediate replacement, even when sufficient funds are available. This lag between identifying changing needs and building new infrastructure results in increased total public costs with respect to design projections. Interim adaptive strategies can be considered, however, to minimize the monetary, environmental, and safety impacts of a sub-optimal design still in the middle of its useful life until it is time to replace the facility [12]. Adapting existing facilities to new use patterns and goals also provides a bridge between generational shifts in infrastructure planning objectives.

In the case of transportation, planning has historically focused on mobility, with the outcomes of ever-increasing VKT and sprawling development patterns that discourage alternatives to the private automobile [13, 14]. In the near term, however, adapting existing facilities to maximize mobility in light of an increased bicycle and transit mode share is an appropriate measure to maximize the value provided by these facilities.

Adaptive strategies comply with existing system constraints and involve lower cost measures to capture the remaining value of past infrastructure investments until such time as a major change consistent with a lower impact vision is warranted. Evaluation of these transitional actions is complicated, however, given the absence of any sort of steady state and the number of analyses that must be integrated.

Most of the individual elements constraining ground transportation systems have been studied in isolation, but few examples of integrated multi-criterion analysis of roadway use have been published. Pavement management systems (PMS) have been developed to help maintenance managers maintain large systems of pavements under budget constraints [15–18]. Separately, traffic engineers have developed microsimulation tools of vehicle dynamics to understand the effect of different road configurations [19–21] and work-zone traffic management [10]. Economic analyses rely in part on the engineering analysis of road utilization and make the connection between more infrastructure and induced economic activity. The interests of pedestrians and bikers are also considered in the context of safety [22–24] and congestion [25] but rarely in terms of providing viable alternatives to automobile transportation.

Efforts to identify significant reductions in the environmental burden of transportation will need to consider these tools together to inform optimal use of roadways under multiple constraints. Here we present a method for combining these analyses with conventional LCA of roadways. We consider the results in the context of adaptive roadway lane (re)configuration, such as converting curb parking to bicycle facilities, that many cities, including Washington, DC and New York City, are currently undertaking in an effort to reduce congestion directly and indirectly by supporting alternatives to automobile travel that make more efficient use of the public right of way and incentivize reduced trip distances.

6.2 METHODS

A total cost minimization approach is proposed in order to identify preferable lane configurations for two-lane urban roadways, given the physical parameters of the site, available right of way width, and traffic volumes by mode. A lane configuration is defined by the number and width (or presence) of parking, bicycle, and conventional travel lanes for each direction. Costs include: annualized pavement maintenance, motor vehicle fuel costs [26], and travel time at half the prevailing wage rate [27]. GHG emissions are also calculated. Computationally, the proposed framework is structured as a series of distinct codes.

Microsimulation of idealized roadway segments was carried out using VISSIM 5.4 for all parameter combinations given in table 1 using common values in table 2. The first section of table 1 lists the parameters that define a lane configuration. The remaining parameters in table 1, taken together, will be referred to as the scenario. The key dynamic explored in this work is the use of a single lane by motor vehicles and bicycles simultaneously, and whether the former is able to safely pass the latter within the lane. VISSIM is able to model lateral behavior within lanes, in addition to more conventional vehicle following and lane changing behaviors, and so is able to consider this question. Peak and off peak traffic volumes were simulated separately and combined using 12 h of each to arrive at daily totals, which were then inflated to annual values. Additional detail is provided in the supporting information (available at stacks.iop. org/ERL/8/015028/mmedia).

TABLE 1: Discrete segment parameter space. All unique combinations were evaluated using microsimulation. The first group defines a lane configuration and the second a scenario.

Travel lane width	3.4 m, 4.3 m
Bicycle lane	None, 1.22 m[1]
Parking lane	None, 2.5
Characteristic length between passing zones	50 m, 100 m, 200 m
Grade	0%, ±4%
Bicycle mode share	1%, 10%

[a] *AASHTO guidelines call for a wider 1.52 m bicycle lane adjacent to curb parking which are applied here as appropriate.*

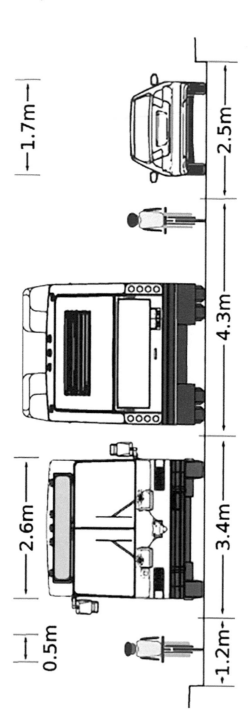

FIGURE 1: A reference multi-modal street section and the relative widths of vehicles and lanes.

TABLE 2: Common parameters and descriptive values for all simulated configurations.

Total width	6.8–16.6 m
Motorized speed	16.7 m s−1
Bicycle speed	Varies by grade
Car occupancy	1.2
Bus occupancy	20
Bicycle occupancy	1.0
Potential curb parking	6.1 m spaces covering 80% of segment length

A low and high bicycle mode share were explored, with 1% representing typical urban mode share in the US and 10% representing a target that leading cities, such as Portland, OR, could achieve in the next decade with sufficient investment. Total person trips are constant between the bicycle volume scenarios, with the difference made up by automobiles with an occupancy of 1.2. Figure 1 shows the relative sizes of vehicles and lanes considered in this work. The design standard for lane width in the US is 3.6 m [28], however it is both permissible and likely that narrower lanes are used in the width constrained urban corridors considered in this study, so we assume a 3.4 m base case. Assuming a 1 m passing buffer, 1.75 m wide automobile, and 0.5 m wide bicycle, an automobile is able to pass a bicycle within the lane, however a 2.6 m wide transit bus is not. To evaluate the benefit of alleviating this "stuck" condition, a wider lane is also considered that allows buses to pass cyclists without either vehicle departing the lane. "Dooring" accidents, crashes between bicycles and opening automobile doors, may also justify additional buffer width between parked cars and lanes used by bicycles, however the present work does not model crashes, so this effect is not represented in the analysis. The impact of these passing conditions is dependent upon the likelihood that buses will encounter a bicycle and become stuck behind it, thereby delaying themselves and following motor vehicles.

Realistic urban corridors vary in width, so it is overly conservative to assume that a given lane width will restrict passing movements indefinitely. Here, we adopt the concept of a characteristic length between passing zones, as given in table 1 to determine the likelihood that a bicycle and bus will be present and the delay expected to result from the encounter. VIS-

SIM simulates these interactions directly. Since Poisson vehicle arrivals are assumed, however, in order to make the results as general as possible, a probabilistic analysis can be carried out to compute expected delay. This analysis is presented in the supporting information (available at stacks.iop. org/ERL/8/015028/mmedia) and agrees with the results of the microsimulation. Motor vehicle speeds were assumed constant for each lane width given the considerable variation that exists in the literature on the effect of lane width of motorist speed choice [22, 29, 28], however, this behavior could be readily altered in the microsimulation parameters where local data is available. Bicycle speeds for each grade were computed according to first principles formulas [30].

Vehicle fuel use and emissions are affected by pavement roughness though not consistently between various operational regimes [31–33] due to the varying contribution of rolling resistance to required power. A power-based vehicle emissions model, CMEM [34], was used to post-process the microsimulation vehicle data at 1 Hz and two roughnesses using a lookup table computed at a reference international roughness index (IRI) of 1.0 m km−1, and for a rough case with an IRI of 4.0 m km−1 by inflating vehicle rolling resistance after Karlsson et al [33]. Only automobile emissions were affected due to the inconclusive results of that study for heavy vehicles. Final emissions and fuel consumption were computed by the pavement management module through linear interpolation of the two roughness cases.

Pavement management activity and emissions were computed based on the previous work of the authors [35] with additional dynamic pavement loading due to roughness [36]. Explicit treatment of heavy vehicles is important in the comprehensive analysis of a roadway given their disproportionate impact [37]. Pavement maintenance plans were computed for both directions of travel lanes, bicycle lanes, and parking lanes independently using aggregated annual vehicle volumes and emissions from the appropriate microsimulation trials. A network Pareto front for all the lanes was then computed, with a representative example given in figure 2. For this analysis, the non-dominated plan with the minimum total GHG emissions, subject to agency constraints, was selected. Further detail on this method, and the larger issues of discounting and temporal variation, can be found in the supporting information (available at stacks.iop.org/ERL/8/015028/mmedia).

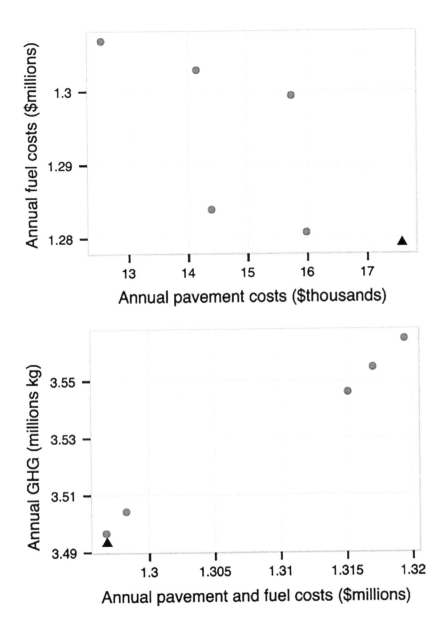

FIGURE 2: Agency pavement costs versus user fuel costs as a result of pavement roughness (upper) and combined costs versus combined GHG emissions (lower) from non-dominated pavement management plans. The triangle data point indicates the selected plan by minimum GHG.

In combining the separate models of pavement management, vehicle microsimulation, and vehicle emissions computation, a hierarchy exists according to the sensitivity of one to another. Pavement roughness influences vehicle fuel use and emissions as well as dynamic pavement loading. Loading affects pavement durability, and maintenance investment determines the resulting pavement condition. Vehicle behavior is assumed to be insensitive to pavement condition within the specified limits, however, which makes travel time cost insensitive to changes in fuel and agency costs. Critically, this means that the PMS optimization can be performed after microsimulation. Otherwise, the task would be computationally intractable with existing microsimulation tools, since a microsimulation would have to be run at each iteration of the PMS genetic algorithm.

A Pareto front of lane configurations can be identified for each scenario that is non-dominated with respect to total costs and width. Lane configurations that include curb parking will naturally not appear in this set since they incur pavement maintenance costs but provide no counterbalancing benefit as computed. A spatial opportunity cost of curb parking can be computed, however, by computing the cost differential between configurations with parking and a point interpolated on the Pareto cost curve at the same total width. This opportunity cost allows decision makers to quantify the potential mobility value of public right of way allocated to parking.

6.3 RESULTS

The effect of bicycle mode share on average travel time for a particular segment can be significant for specific conditions as shown by the results in figure 3. Differentiation between the cases occurs when heavy (wide) vehicles, such as transit buses encounter a bicycle and have insufficient room to safely pass resulting in significant delays for themselves and the motor vehicles behind them. In these graphs, individual data points represent microsimulations of specific cases and are plotted with random jitter on the length axis for legibility. As with other results presented here, they are normalized to a kilometer of travel. The trend lines are second order polynomials used to illustrate the relationships of interest. The effect of grade, and whether or not vehicles are traveling uphill or down, have important effects on the results presented in figure 3. The data are grouped based on whether or not buses are stuck behind bikers in the different configurations. For the level ground segment, the results are equivalent. For the inclined segment, the uphill travel time is always considerably higher than downhill travel time if trucks get stuck behind buses. The impact of the stuck condition is proportional to the relative speed difference between bicycles and motor vehicles, which comes from roadway grade, and the likelihood of a heavy vehicle encountering a bicycle within the characteristic distance between passing zones. This is determined by modal volumes and headway distribution.

For lane widths more narrow than those considered here, all motor vehicles with more than two wheels would be unable to pass bicycles within their lane, with the result that the expected speed of all traffic would approach that of bicycles as characteristic length and bicycle volume increased. These cases are not presented in order to focus on the more typical but less intuitive stuck condition, and because very narrow lanes are likely to have an effect on driver speed decisions according to the particular characteristics of the site, such as sight distance, land use, number of driveways, and other factors. This is not to say, however, that the framework presented here is not suitable for 3.1 m lane widths, only that the results would not be transferable to other situations. For wider lane widths, the differentiation observed here disappears as heavy vehicles are able to pass bicycles. All four stuck groups include multiple lane configurations which reveal more subtle differences in annual costs with respect to width (distance between curbs) when considered individually. In figure 4, this relationship is seen as a distinct Pareto optimal frontier for each characteristic length between passing zones, grade, and bicycle mode share considered, with non-dominated lane configurations shown in bold. The influence of the stuck condition can be seen in the abrupt transition in the Pareto curve around 8.6 m, or the minimum width of a configuration not stuck in both directions. The initial drop between the first two non-dominated points is larger for the 4% case, as compared to level ground, since alleviating the stuck condition on the uphill segment is considerably more important than in the downhill direction given the dramatic difference in expected bicycle speeds.

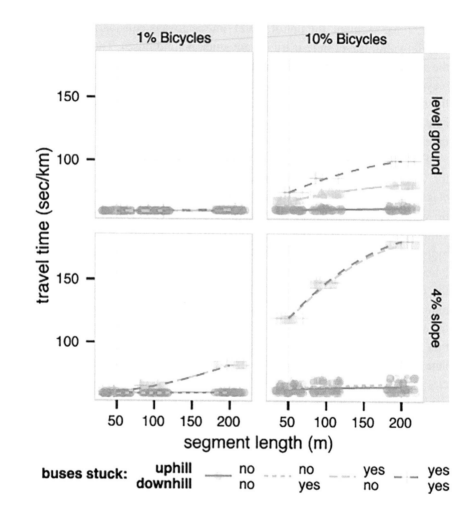

FIGURE 3: Simulated motorized travel time illustrates the delay caused when heavy vehicles are unable to pass bicycles.

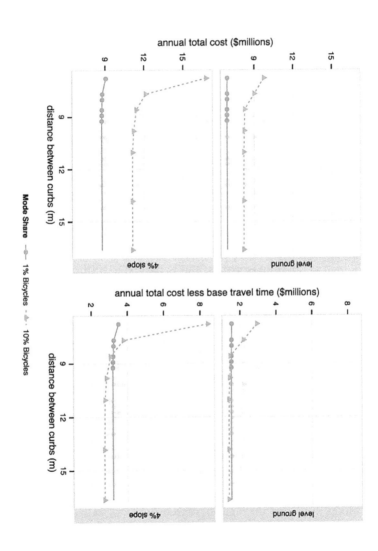

FIGURE 4: The relationship between road width and costs follows a Pareto optimal behavior, shown here for 100 m characteristic lengths. Total travel time costs (left) dominate fuel and PMS costs, however when delay costs are used (right) increased cycle mode share can reduce costs for appropriate lane configurations.

For the urban situations considered in this work, both travel time and fuel use are positively correlated with vehicle delay. These costs are also considerably larger than agency expenditures for non-dominated pavement maintenance plans, which can be seen in figure 2. This relationship supports the decision to select the PMS plan with the minimum total GHG emissions (higher agency cost), since a comparatively small agency investment provides a larger reduction in user costs. It follows that minimizing total costs also minimizes GHG emissions since total costs are dominated by time and fuel and are sensitive only to delay once pavement condition has been established. This is encouraging since typical planning processes do not explicitly quantify GHG emissions.

Figure 4 presents the Pareto front with respect to total costs and then using delay rather than total travel time costs. Neither approach is strictly more accurate, however since bicycle travel is generally more time consuming over the same roadway segment, increasing bicycle mode share tends to dramatically increase time costs. Using the time cost of delay only, assumes that travelers had already accounted for this cost externally, which is not unreasonable, and is typical in traffic analysis. Under this assumption, figure 4 reveals a tipping point where increased bicycle mode share lowers total costs, given sufficient roadway width. For cases where significant increases in bike ridership are not paired with enhanced facilities like wider roads, this will increase travel time for all users.

Parking is not valued in the total cost reported in figure 4. As a result, lane configurations with parking bays incur pavement maintenance costs without corresponding negative costs from parking's value as a service, rendering these configurations sub-optimal. This also explains why costs reach their minimum in figure 4 approximately 5 m before the maximum width since that is the width occupied by two 2.5 m parking bays. Curb parking does have a site-specific value both in terms of vehicular accessibility and in the broader economic sense by supporting value capture along the street through increased business patronage. Because the magnitude of this value capture is so site-specific, figure 5 presents a spatial opportunity cost of curb parking as the difference between the total cost of a lane configuration that includes parking and a linear interpolation of the Pareto front at the same width. Visually, this is the vertical distance between the lighter data points in figure 4 and the Pareto front, normalized to a daily value per parking space. The narrowest lane configuration that can include parking is 9.3 m wide, so opportunity cost is reported from this value up to the maximum configuration width.

A spatial opportunity cost of parking is shown in figure 5 for the inclined case with a 100 m characteristic distance between segments. The likelihood of buses' getting stuck behind a bicycle is related to characteristic length, so other scenarios look similar to this plot but shifted in magnitude accordingly by length. Level ground also exhibits consistent behavior, albeit with a different curve shape that can be inferred from figure 4. Since the maximum hourly volumes considered in this study are below saturation levels, delay comes almost entirely from the bicycle/motor vehicle interactions. Therefore, it is not surprising that parking opportunity costs are highly sensitive to the percentage of bicycles in the traffic stream. For a given width, parking and bicycle infrastructure essentially compete for the same space which results in large opportunity costs for parking in narrow configurations, as might be seen in a traditional urban neighborhood of places like New York City and Washington, DC. These larger values are also well in excess of typical parking meter returns, which is consistent with the pervasive subsidization of automobile parking in the US [38]. These values must also be considered conservative since they do not include any opportunity costs for land in the pedestrian zone on either side of the roadway proper that could be put to other uses, such as pedestrian mobility or restaurant seating.

6.4 SENSITIVITY AND ADDITIONAL CONSIDERATIONS

To develop a more complete understanding of how urban roadway design can be informed by a model like this, it is useful to consider both those factors that had little impact on the results and those factors that were not included in the analysis for one reason or another. Pavement roughness and its effects on fuel consumption are considered here even though in the context of urban driving, roughness is far less important than acceleration cycles in determining fuel efficiency. However, since travel time is independent of roughness, pavement management costs for a given configuration are optimized against marginal vehicle fuel consumption which is on the same order of magnitude. Additionally, the fiscal reality of the pavement manager may be such that maintaining serviceable pavement is a constant struggle, and so minimization of agency costs may replace minimization of GHG emissions in selecting the best plan. The impact pavement condition has on a given commuter's likelihood to select the bicycle or automobile mode is difficult to quantify, but also more significant to the ultimate makeup of the Pareto set of lane configurations by determining relative volumes between the modes as can be seen in figures 4 and 5.

Similarly, only select measures of the health and safety factors associated with modal shift are captured by this model since many of these are difficult to quantify with confidence even though they are often cited as an important driver and/or obstacle in mode shift toward more active forms of transportation [39]. Crashes, though not considered in this work, are an important factor in selecting bicycle transport and even though AASHTO guidelines dictate an additional 0.3 m of width for bicycle lanes adjacent to curb parking, that width remains insufficient to prevent dooring crashes between bicycles and parked cars [40, 41]. The cost of curbside parking goes up fractionally for this additional 0.3 m of pavement to address the risk of dooring to bikers, however the actuarial cost of a potential fatality is on the same order of magnitude as total annual travel cost per kilometer in this study. Crashes are difficult to model generally, and difficult to even estimate based on past reports involving bicycles due to pervasive under-reporting of non-fatal encounters. Never the less, the potential health and safety costs are potentially large enough to influence the results presented here.

Another limitation of the model is that it does not consider roadway segments within the larger roadway network. In many urban areas, vehicle arrival is not Poisson distributed but rather appears as a decaying platoon progressing between controlled intersections. In addition, driveways play an important roll in platoon migration and overall capacity in urban settings. Finally, buses stop periodically for passengers and this presents an opportunity for the traffic to clear and the biker and bus to separate. These factors were not considered here in order to make the analysis as general as possible but could be readily incorporated into the microsimulation for a specific site. Many of these factors would need to be included in a similar model to derive site-specific estimates for the opportunity cost of parking even though we expect that the general trends discussed here would hold for most urban roadways.

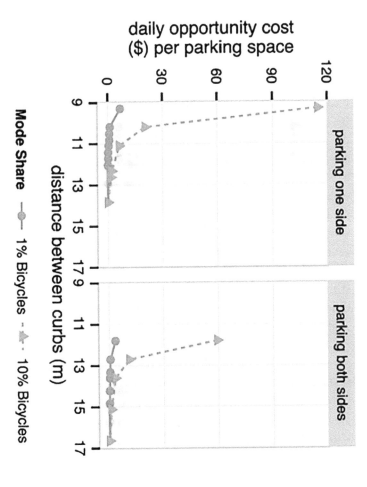

FIGURE 5: Spatial opportunity cost for curb parking, shown here for 100 m characteristic length and 4% grade, increases dramatically for narrower roads suggesting that when all the costs are considered, there is a tipping point beyond which curb parking becomes an expensive use of land

6.5 IMPLICATIONS

A major challenge for urban areas around the world is to improve livability, which is often achieved by reducing reliance on automobile transport [42, 43]. A shift away from the automobile also results in significant reductions in energy use and GHG emissions, both of which are increasingly relevant policy objectives. For countries with a legacy of automobility-dominated planning and policy decisions, such as the US, these goals are especially daunting. Reducing VKT through more compact development and alternative transportation modes is a long-term objective, but previous studies have observed that many of our projected future emissions are "locked in" by virtue of the inefficient nature of our existing infrastructure stock and the relatively slow rate at which this is replaced [44]. Short-term strategies are needed to achieve some of these gains without major changes to our existing infrastructure. Improving bicycle facilities is one such change that could encourage higher mode share without significant investment. In cities like Washington, DC and New York City, where many of the trips are short and well suited to bicycle transport, efforts have been underway for several years to provide such facilities. Consequently, bike mode share is increasing significantly [45], but the ridership rates in these US cities are still an order of magnitude lower than in many European cities [46]. Additionally, these changes are not without controversy in light of their direct costs and use of limited space [47].

Tools like the one developed here can be used to help resolve the apparent conflicts that could inhibit progress toward more environmentally sustainable infrastructure. We find that increasing bicycle mode share can have a significant impact on motor vehicle delay, and indeed greatly increase total costs where sufficient right of way is not provided. Conversely, with sufficient space to allow wide vehicles to pass bicycles, a reduction in total costs for all users is obtained through an increased bicycle mode share. Parking, specifically curb parking, emerges as space that is potentially most eligible for reallocation to bicycles in width constrained urban corridors because of its ideal position and the relatively minimal expense of lane reconfiguration. The spatial opportunity cost of parking quantifies this tradeoff and exhibits sharp transitions between realistic monetary amounts for wider streets and unfeasibly high expected returns for more narrow areas. These results provide a clear guide to traffic engineers and urban policy makers with respect to optimal allocation of limited pavement.

Even though life cycle assessment and other tools are useful for beginning to understand pieces of this problem, a systems-based approach like the one proposed here is needed to directly support policy decisions. By considering infrastructure systems in an integrated, yet quantitative manner using existing modeling frameworks, short-term low cost opportunities emerge for efficiency improvements that would not be obvious using other tools alone. The general framework proposed here could be applied in a variety of infrastructure contexts. For example, the green building industry standard, leadership in energy and environmental design (LEED), has already recognized that buildings cannot be analyzed outside of the larger infrastructure context within which they exist and is making changes to consider the community within which the building exists [48]. Other types of infrastructure have been less studied and offer hereto untapped efficiency improvements. Water treatment and distribution, freight transport, the electrical grid, and wireless communications, and others could all benefit from considering engineering performance criteria, agency and user cost, and human factors together for informing improved policy.

REFERENCES

1. Wilson J H, Williams L Z, Schreiber J J, Mullen M A, Peterson T D and Strait R 2009 State approaches to reducing transportation sector greenhouse gas emissions Transportation, Land Use, Planning, and Air Quality: Selected Papers of the Transportation, Land Use, Planning, and Air Quality Conference 2009 (Reston, VA: ASCE) p 9
2. Keoleian G A, Kar K, Manion M M and Bulkley J W 1997 Industrial ecology of the automobile: a life cycle perspective Technical Report (Warrendale, PA: Society of Automotive Engineers (SAE))
3. Hendrickson C T, Lave L B and Matthews H S 2006 Environmental Life Cycle Assessment of Goods and Services: An Input–Output Approach (Washington, DC: Resources for the Future)
4. Thenoux G, Lvaro G and Dowling R 2007 Energy consumption comparison for different asphalt pavements rehabilitation techniques used in Chile Resour. Cons. Recy. 49 325–39

5. Horvath A 2003 Life-cycle environmental and economic assessment of using recycled materials for asphalt pavements Technical Report 510 (Berkely, CA: University of California Press)

6. Hendrickson C, Horvath A, Joshi S, Lave L and Design G 1998 Economic input–output models for environmental life-cycle assessment Environ. Sci. Technol. 32 184

7. International Energy Agency 2009 Transport Energy and CO2 (Paris: IEA Publications)

8. Santero N J and Horvath A 2009 Global warming potential of pavements Environ. Res. Lett. 4 034011

9. Norman J, MacLean H L and Kennedy C A 2006 Comparing high and low residential density: life-cycle analysis of energy use and greenhouse gas emissions J. Urban Plann. Dev. 132 10–21

10. Huang Y, Bird R and Bell M 2009 A comparative study of the emission by road maintenance works and the disrupted traffic using life cycle assessment and micro-simulation Transp. Res. D 14 197–204

11. Golob T F, Recker W W and Alvarez V M 2004 Freeway safety as a function of traffic flow Accident Anal. Prevent. 36 933–46

12. Pawlovich M D, Li W, Carriquiry A and Welch T 2006 Iowa's experience with road diet measures: use of Bayesian approach to assess impacts on crash frequencies and crash rates Transp. Res. Record, J. Transp. Res. Board 1953 163–71

13. Ewing R 1993 Transportation service standards—as if people matter Transp. Res. Record, J. Transp. Res. Board 1400 10–7

14. Miller J S and Evans L D 2011 Divergence of potential state-level performance measures to assess transportation and land use coordination J. Transp. Land Use 4 81–103

15. Gu W, Ouyang Y and Madanat S 2012 Joint optimization of pavement maintenance and resurfacing planning Transp. Res. B 46 511–9

16. Lidicker J, Sathaye N, Madanat S and Horvath A 2012 Pavement resurfacing policy for minimization of life-cycle costs and greenhouse gas emissions J. Infrastruct. Syst. at press (doi:10.1061/(ASCE)IS.1943-555X.0000114)

17. Zhang H, Keoleian G A and Lepech M D 2012 Network-level pavement asset management system integrated with life cycle analysis and life cycle optimization J. Infrastruct. Syst. 19 99–107

18. Sathaye N and Madanat S 2012 A bottom-up optimal pavement resurfacing solution approach for large-scale networks Transp. Res. B 46 520–8

19. Fellendorf M 1994 VISSIM: a microscopic simulation tool to evaluate actuated signal control including bus priority 64th Institute of Transportation Engineers Annual Mtg

20. Faghri A and Egyhaziova E 1999 Development of a computer simulation model of mixed motor vehicle and bicycle traffic on an urban road network Transp. Res. Record, J. Transport Res. Board 1674 86–93

21. Behrisch M, Bieker L, Erdmann J and Krajzewicz D 2011 SUMO—simulation of urban mobility: an overview SIMUL 2011: 3rd International Conf. on Advances in System Simulation (Barcelona, Oct. 2011) pp 63–8

22. Noland R B 2003 Traffic fatalities and injuries: the effect of changes in infrastructure and other trends Accident, Anal. Prev. 35 599–611
23. Johan de Hartog J, Boogaard H, Nijland H and Hoek G 2010 Do the health benefits of cycling outweigh the risks? Environ. Health Perspect. 118 1109–16
24. Persaud B and Lyon C 2010 Evaluation of lane reduction 'Road Diet' measures on crashes and injuries Technical Report 1 (Washington, DC: Federal Highway Administration)
25. Dowling R, Flannery A, Ryus P, Petrisch T and Rouphail N 2010 Field test results of the multimodal level of service analysis for urban streets National Cooperative Highway Research Program Web-Only Document 158 (Washington, DC: Transportation Research Board)
26. American Automobile Association 2007 Fuel Gauge Report
27. AASHTO 2010 User and Non-User Benefit Analysis for Highways (Washington, DC: American Association of State Highway and Transportation Officials)
28. Transportation Research Board 2010 Highway Capacity Manual (Washington, DC: Transportation Research Board of the National Academies)
29. Rosey F, Auberlet J-M, Moisan O and Dupre G 2009 Impact of narrower lane width: comparison between fixed-base simulator and real data Transp. Res. Record, J. Transp. Res. Board 2138 112–9
30. Zorn W 2008 Speed and Power Calculator (www.kreuzotter.de/english/espeed.htm, referenced 20 October 2012)
31. Barnes G and Langworthy P 2003 The per-mile costs of operating automobiles and trucks Technical Report (St Paul, MN: Minnesota Department of Transportation)
32. Zaabar I and Chatti K 2010 Calibration of HDM-4 models for estimating the effect of pavement roughness on fuel consumption for US conditions Transp. Res. Record, J. Transp. Res. Board 3 105–16
33. Karlsson R, Hammarström U, Sörensen H and Eriksson O 2011 Road surface influence on rolling resistance Coastdown measurements for a car and an HGV Technical Report
34. Barth M, An F, Younglove T, Scora G, Levine C, Ross M and Wenzel T 2000 NCHRP project 25–11 development of a comprehensive modal emissions model National Cooperative Highway Research Program Web-Only Document 122 (Washington, DC: Transportation Research Board)
35. Gosse C, Smith B and Clarens A 2012 Environmentally preferable pavement management systems J. Infrastruct. Syst. at press (doi:10.1061/(ASCE)IS.1943-555X.0000118)
36. Fekpe E 2006 Pavment damage from transit buses and motor coaches Technical Report (Columbus, OH: Batelle Memorial Instititue)/
37. Diefenderfer B K, Moruza A K, Brown M C, Roosevelt D S, Andersen E C, O'Leary A A and Gomez J P 2010 Development of a weight-distance permit fee methodology for overweight trucks in Virginia Int. J. Pavement Res. Technol. 2 236–41
38. Shoup D 2011 The High Cost of Free Parking 2nd edn (Washington, DC: American Planning Association)
39. Jacobsen P L, Racioppi F and Rutter H 2009 Who owns the roads? How motorised traffic discourages walking and bicycling Injury Prev.: J. Int. Soc. Child Adolescent Injury Prev. 15 369–73

40. 2006 Bicyclist fatalities and serious injuries in New York City 1996–2005 Technical Report (New York: New York City Departments of Health and Mental Hygiene, Parks and Recreation, Transportation, and the New York City Police Department)

41. Works Department and Emergency Services 2003 City of Toronto bicycle/motor-vehicle collision study Technical Report (Toronto: Transportation Services Division Transportation Infrastructure Management Section)

42. Noland R B and Lem L L 2002 A review of the evidence for induced travel and changes in transportation and environmental policy in the US and the UK Transp. Res. D 7 1–26

43. Rahul T M and Verma A 2013 Economic impact of non-motorized transportation in Indian cities Res. Transp. Econ. 38 22–34

44. Lemer A C 1996 Infrastructure obsolescence and design service life J. Infrastruct. Syst. 2 153–61

45. 2011 NYC commuter cycling indicator Technical Report (New York: New York City Department of Transportation)

46. 2005 District of Columbia bicycle master plan Technical Report (Washington, DC: District Department of Transportation)

47. Krizec K 2007 Estimating the economic benefits of bicycling and bicycle facilities: an interpretive review and proposed methods Essays on Transport Economics (Contributions to Economics) ed P Coto-Millán and V Inglada (Heidelberg: Physica-Verlag) pp 219–48

48. US Green Building Council 2009 LEED Reference Guide for Green Neighborhood Development LEED 2009 edn (Washington, DC: US Green Building Council)

PART III

AVIATION

Sustainable Development and Airport Surface Access: The Role of Technological Innovation and Behavioral Change

TIM RYLEY, JAAFAR ELMIRGHANI, TOM BUDD,
CHIKAGE MIYOSHI, KEITH MASON RICHARD MOXON,
IMAD AHMED, BILAL QAZI, AND ALBERTO ZANNI

7.1 INTRODUCTION

Sustainable development reflects an underlying tension between achieving economic growth and addressing environmental challenges, and this is particularly the case for the aviation sector. Although the sector is largely considered to be economically and socially sustainable, it also generates environmental concerns because of climate change impacts from aviation-related emissions. Despite a dip due to the current economic recession, United Kingdom air travel has increased over the last ten years. There were 219 million terminal passengers at UK airports in 2011 compared

Sustainable Development and Airport Surface Access: The Role of Technological Innovation and Behavioral Change. © Ryley T, Elmirghani J, Budd T, Miyoshi C, Mason K, Moxon R, Ahmed I, Qazi B, and Zanni A. Sustainability 5,4 (2013), doi:10.3390/su5041617. Licensed under Creative Commons Attribution 3.0 Unported License, http://creativecommons.org/licenses/by/3.0/.

with 167 million in 1999 [1]. There is also likely to be a further long-term growth in demand with a knock-on impact on emissions such as carbon dioxide (CO_2). Although much of the aviation-related focus has fallen on reducing aircraft emissions, airports have been under increasing pressure to support the vision of a low carbon energy future. In particular, in recent years there has been a focus on reducing the share of emissions from surface access journeys to and from the airport.

This paper contains an initial review of the issues surrounding sustainable development and airport surface access. It focuses on two aspects: an evaluation of the technological innovation options that will enable sustainable transport solutions for surface access trips, and a discussion of the role of behavioral change for these journeys from a theoretical perspective using empirical data from Manchester airport. Finally, the potential contribution of technology and behavioral intervention measures to improvements in sustainable development relating to surface access is discussed.

This paper presents findings from one of a series of Airport Operations projects (funded by the United Kingdom Research Councils' Energy Programme), the 'ABC project: Airports and Behavioural Change: towards environmental surface access travel'. The project aims to encourage better environmental behavior of individuals travelling to and from airports (the surface access component of air travel), and has a focus on sustainable transport solutions for the year 2020, a mid-term timescale. A unique aspect of the ABC project is that is brings together two components, surface transport and air travel, as each transport component has environmental imperatives to reduce both travel demand and carbon emissions.

7.1.1 SUSTAINABLE DEVELOPMENT AND AIRPORT SURFACE ACCESS

Climate change has had an increased role over time within the environmental aspects of sustainable development, as shown by its prominence within the 2005 UK Sustainable Development Strategy [2]. Transport is a major contributor to greenhouse and pollutant emissions, and transport is one of the only sectors where emissions have been increasing [3]. This is especially the case for aviation. While it is estimated that commercial air travel currently accounts for around 2% of global CO_2 emissions [4], it is expected that this figure will rise given the projected growth of the sector in the future.

The UK Government's commitment to reduce CO_2 emissions by 80% by 2050 over 1990 levels, with an interim target of a 34% reduction by 2020 [5], has put the issue of aviation related emissions into focus. Given the projected growth in the sector, it is likely that aviation will take an increasingly significant proportion of any carbon budget [6]. The UK Government forecasts growth in UK aviation, hence doubling carbon emissions from approximately 9 million tonnes of carbon (MtC) in 2000 to 17.4 MtC in 2050 [7].

It should be noted that there are many significant environmental impacts of air transport including, amongst others, the development of airports and associated infrastructure; noise and vibration from aircraft (and surface access); water pollution (e.g., surface run-off); local air quality pollutants (e.g., CO, NOX); solid waste (scrapped aircraft, waste oil/tires); other waste for disposal; energy and water consumption; and complex land-based supply chain operations. This paper considers the role of the airport developing in a sustainable manner. One of the immediate necessities to initiate this process by conducting research which helps understand the key challenges facing airports, and to facilitate the development of solutions. A challenging objective for airports is to develop in a sustainable manner, increase airport capacity and economic performance, while simultaneously minimizing environmental impact.

While surface access emissions are relatively low in comparison to those from aircraft, they are one of the primary sources of emissions that airports have the ability to influence. Hence this paper, and the underlying research project from which it stems, focuses on surface access journeys to and from airports.

Access to airports is an essential part of airport operations as well as being of particular importance for travellers. It is estimated that 65% of journeys to large airports in Europe and the US are made by private cars, with this figures rising up to 99% for smaller regional airports [8,9]. Importantly, previous studies have demonstrated that the Value of Access Time is considerably higher than the estimated value of time spent travelling normally for commuters e.g., [10]. The reason for this appears to be the risk of missing flights, which increases when travel time to airport increases [11]. In addition, it is important to analyze the arrival time of passengers and their respective flight departure times. This is estimated to depend, in particular, on whether: 1) the passenger is flying for business or leisure, with business travellers preferring to arrive later at airport; 2) the passenger is in employment or retired, the frequency of flights in the past; and 3) the passenger is travelling with any luggage or not [10]. On the other hand, airport managers face a difficult task with regards to reconciling congestion and environmental pressures to reduce private vehicle trips with the substantial commercial importance of car parking revenues [12].

Interviews with a range of surface managers at an early stage of the ABC project revealed a wide variety of surface access issues and management policies [13]. The need to reduce the share of passenger journeys made by private car and to increase public transport use was identified as a key issue, with a particular focus on reducing 'drop-off/pick-up' journeys. However, it was also found that while reducing private car journeys may yield environmental benefits, such strategies are largely at odds with substantial commercial pressures to maximize the revenue potential of airport parking. Particular focus was paid to the problem of changing current surface access travel behavior among airport users. UK airport surface access managers are reliant on a range of external stakeholders, whom the airport has little direct control over, such as operators of key infrastructure and train companies. Whilst in many cases the airport—stakeholder relationship is shown to be mutually beneficial, airports are still in a vulnerable position.

Surface access to airports has a key aspect of reducing access by motorized transport, particularly where alternative modes could easily be utilized. Surface access modes under consideration include a range of public transport options such as taxi, bus and rail. The travel behavior issues for surface access are different from other transport contexts; for instance, when accessing airports individuals will often not use public transport as they have to carry luggage with them, and hence the need for sufficient luggage storage capacity on board public transport that is visible throughout the journey, or, where appropriate, off-site luggage drop-off facilities.

The types of people who use airports can be seen primarily as passengers and employees. For passengers, there is a particular concern with the drop-off/pick-up of air travellers, typically by a family member or friend. This type of journey, together with those by taxi, is twice the number of vehicle trips than by a passenger who travels by car and parks at the airport. Where possible, airports will try to reduce these trips in favor of driving and parking and, ultimately, public transport. Employee surface access issues can vary considerably from those for passengers. Generally, there is very high reliance on private car journeys for airport employees, who typically need to access the airport regularly, reliably, cost effectively and at times of the day that are not always well served by public transport. In the UK, airport employees typically have their parking subsidized or paid for by their employer. Figure 1 shows a hierarchy of preferred surface access modes for passengers, from the least environmentally sustainable of drop-off / pick-up to the most environmentally sustainable of public transport. It has been adapted from the Manchester Airport Surface Access Strategy [14].

7.1.2 SURFACE ACCESS AT MANCHESTER AIRPORT

Initial work included qualitative interviews at Manchester Airport (together with some at Robin Hood Doncaster Sheffield Airport). The interviews showed that passengers most likely to drop-off/pick-up, or drive and park, are: holidaymakers, those in groups and/or those with a lot of luggage. Passengers and employees also mentioned the lack of public transport in the early mornings. There has also been some analysis of Manchester airport secondary data from the Civil Aviation Authority (CAA) and employers on the airport site. Figure 2 shows the travel mode share for surface access at Manchester Airport in 2009. It demonstrates the dominance of private transport, which increased from 80% in 1996 to 90% in 2009.

Passenger carbon emissions (grams per passenger km in 2009) from CAA data estimated at Manchester Airport, by transport mode [16], shows:

- Highest emissions were from car users, particularly 'Drop-off / pick-up' (221 g/km – 57% of total emissions) & 'taxi' (229 g/km) passengers;
- Emissions per passenger km of 'car and park' (96 g/km) are lower; and
- Rail (77 g/km) and bus (50 g/km) emissions per passenger are the lowest.

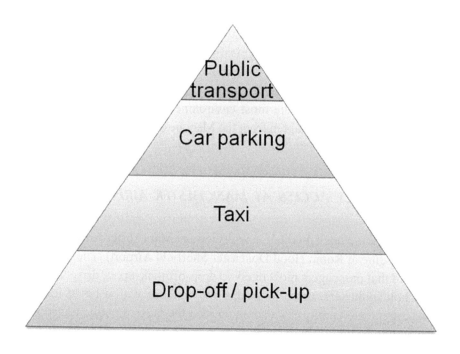

FIGURE 1: A hierarchy of surface access modes in order of environmental sustainability (from public transport the most environmentally sustainable to drop-off / pick-up the least environmentally sustainable) Source: [14] (figure should be centered).

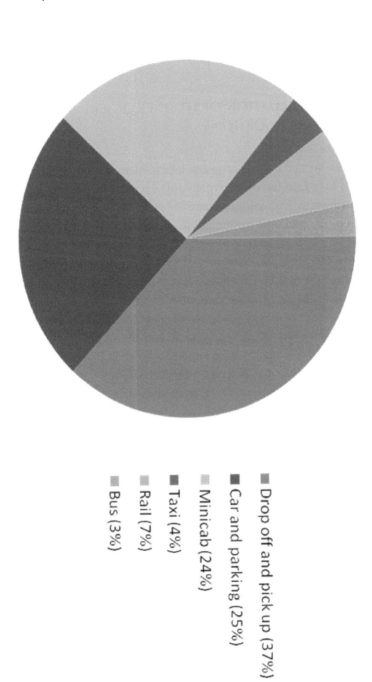

FIGURE 2: Travel mode share for passenger surface access to Manchester Airport in 2009 (source: [15]).

Figures for leisure passengers were lower than for business travellers due to higher load factors per car mode.

7.2 AN EVALUATION OF TECHNOLOGICAL INNOVATION OPTIONS

In this section a range of technological innovation options are evaluated that aim to reduce carbon emissions for airport surface access journeys. The following three technology options are evaluated in turn: telepresence (to reduce drop-off / pick-up), techniques to encourage public transport use (e.g., RFID tagging of luggage) and techniques to encourage sharing rides (e.g., software development). There are technologies to provide home telepresence that can reduce the number of surface access trips that involve drop-off/pick-up of passengers. Communications and software tools can be developed to provide a telepresence experience at home using Internet Protocol, home broadband and home sound surround systems. An illustration of a telepresence system that could be installed at an airport is shown in Figure 3.

Rather than travel to an airport to drop-off and pick-up a passenger, relatives or friends could order an on-demand event to say goodbye to the traveller where telepresence can offer a realistic experience. Remote meeting solutions can reduce carbon emissions by cutting the number of travellers and vehicles' users. Many companies currently use tele-conferencing as an alternative to face-to-face meetings. However, remote meeting solutions can suffer from a lack of natural human sense and network quality of service. Telepresence provides an alternative solution by facilitating real-time connections, face-to-face interactions and lifelike size. The lack of natural human sense is minimized and thus this new technology supports people in learning, play, work and meetings with a high quality of service while they are at different locations. Telepresence offers a more 'three-dimensional' experience than standard television viewing. Although some consumers currently use packages such as Skype and iPad facetime, due to the large number of (potential) customers and low bandwidth of the internet user, such applications fail to provide the required quality of service. Telepresence works on dedicated high bandwidth connections, which means that the level of experience is not comparable.

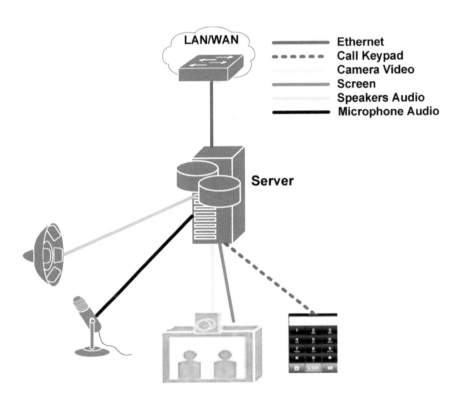

FIGURE 3: Illustrated telepresence system to be installed at an airport.

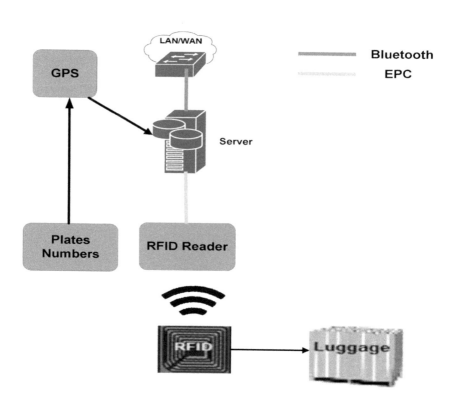

FIGURE 4: Illustrated radio-frequency identification (RFID) system to be installed in trains/coaches.

In addition, relatives or friends at multiple locations could join the event. An increased Internet Protocol television and broadband service penetration is an important enabler for this approach. Airports could offer telepresence suites that include dynamic video, motion sensitive cameras and surround sound, not only for business use but also for the general traveller. Indeed, telepresence is being introduced at some large airports to assist passengers with way-finding see [17]. This demonstrates that it is a feasible application as a substitute for drop-off / pick-up journeys. Telepresence is at the first level of development and application (there are currently several telepresence providers e.g. Cisco), and so over time is likely to have higher market penetration ratio and lower cost. It could, therefore, be a feasible technological application for use at airports in the year 2020.

An important consideration, like other technological innovations, is the extent to which passengers will embrace this particular technology, and their willingness to pay for it. The recent growth of personal video communication via smart phones and tablet computers presents both cause for optimism and caution with this regard. This technology is being investigated within the ABC project in terms of a hypothetical network of base stations within an airport terminal to support telepresence. In an airport environment, passengers act as mobile nodes, moving in random ways and could initiate telepresence. There could be a number of base stations installed to enable high-definition broadband services and good signal coverage for passengers, but this would be at a certain cost.

There are technologies to encourage use of public transport (coach or rail). Given that carrying luggage is a significant barrier that can put off passengers from taking public transport for surface access journeys, a remote check-in system could be of assistance. A remote check-in system would be particularly attractive if passengers were able to track and locate their luggage along the journey. To facilitate luggage tracking and identification RFID (radio-frequency identification) technology can be implemented. An illustration of the RFID system that could be installed in trains and/or coaches is shown in Figure 4.

Continuous information about luggage location would be available through a web interface or text service. When an air passenger gets to the train (or coach) station on a specified travel date, a Bluetooth auto message would provide the train/coach options to synchronize the passenger and their luggage for the surface access journey. There could be a synchronized loading of the luggage onto the train/coach and collection by baggage handlers (or passengers) at the airport. To avoid security problems, passengers and their luggage could travel on the same coach/train. RFID technology could help to minimise difficulties associated with large numbers of passengers and volumes of luggage at the airport. After the luggage drop-off, RFID will enable passengers and staff to keep track of their luggage during the journey to the airport. This type of system has already been deployed at Hong Kong International Airport, where RFID tags on boarding passes and luggage are utilized for safety, security and travel arrangement purposes.

There are technologies to encourage vehicle sharing. It is common for travellers to form groups when using taxis or (mini) buses for surface access journeys. This simple idea could impact upon carbon emissions for surface access trips. When an individual books air travel, there could also be a choice for them to share surface travel to and/or from the airport, perhaps with people not known to them. An alternative (or complementing system) is for the development of a software tool owned by the airport that would send messages (email or text) to other passengers within a given geographic radius of each other and with compatible journey times. Confidential personal information would be removed to establish willingness to share a surface access journey. Optimum scheduling and route selection software would then match passenger requirements and inform the travellers selected.

7.3 UNDERSTANDING BEHAVIORAL CHANGE OPTIONS

It is important to gain a deeper understanding of the factors that determine passenger behavior for journeys to and from airports, and there is a need to examine the relative importance of psychological and situational factors as determinants of mode choice. A questionnaire survey of 860 departing air passengers at Manchester Airport was conducted in June-July 2011 to measure psychological constructs pertaining to two well established theories of attitude-behavior relations, namely the Theory of Planned Behavior [18] and the Norm-Activation Model [19]. In addition, situational variables relating to various aspects of the passenger's trip, background information about their general travel behavior, and socio-demographic information were also elicited.

Psychological behavioral models such as the Theory of Planned Behavior and the Norm-Activation Model have been used extensively in travel behavior research in recent years. A central assumption of the Theory of Planned Behavior is that the concept of behavioral intention is the key antecedent of actual behavior. It assumes that if alternative behaviors exist a choice is made based on the relative strengths of the intentions to perform each alternative [20]. The Norm-Activation Model [19] takes a very different perspective on behavior [21]. While the Theory of Planned Behavior stresses personal-utility, the Norm-Activation Model focuses on the role of personal morals. The central assumption of the Norm-Activation Model is that feelings of personal moral obligation (known as personal norms) are the only causal determinants of behavior [22].

The Theory of Planned Behavior has been shown to be most suited to decisions that are motivated by personal utility maximization, whereas the Norm-Activation Model is more suited to behaviors that contain a moral element. Consequently, mode choice is a decision that can involve both of these factors. A number of mode choice studies have employed a joint model that incorporates elements of both theories.

Initially, the Theory of Planned Behavior and the Norm-Activation Model were tested against the data using structural equation modeling; a statistical technique used to test the structural validity of theoretical models. Results of the analysis indicate that public transport use is determined predominantly by behavioral intentions, as posited in the Theory of Planned Behavior, rather than personal moral obligations, as suggested in the Norm-Activation Model.

This informed the next stage of analysis where two combined models, containing constructs from both the Theory of Planned Behavior and the Norm-Activation Model, were tested. In addition, these models also included constructs relating to descriptive norm (perceptions of what is 'normal' behavior in a population), efficacy (perceptions of what can be achieved) and anticipated feelings of guilt if one were to always use their car to get to the airport instead of using public transport. They were included to see to what extent they improved the predictive power of the models.

Results of the structural equation modeling procedure show that, overall, the combined models are useful determinants of public transport use but the additional constructs do not add to their predictive ability.

It is difficult for behavioral models to fully take into account the multitude of situational and socio-demographic variables that affect behavior. Regarding surface access travel, for example, the purpose of a passenger's trip has been identified as an important influence on mode choice. Business passengers may place a higher value on their time than leisure passengers [23,24], but a lower value on the cost of their trip [25]. Leisure passengers may also be more likely to be carrying heavy luggage with them than business passengers [26], which may affect their choice of mode.

As a result, items in the survey were included to elicit information relating to various situational variables and socio-demographic information. Situational variables included: the purpose of the passenger's journey, the geographical origin of their surface access trip, whether they had started their journey from home, their place of work, how many bags they were carrying with them and the size of their travel party. Information about the number of flights the passenger had taken in the previous twelve months was included, in addition to background information about general travel behavior. Socio-demographic information relating to passenger age, nationality and residence was also collected. One item was included to determine willingness to share a ride to the airport with other passengers in the future. This item was designed specifically to link in with the evaluation of technologies for increasing ride sharing, which was discussed in the previous section. A small but sizeable proportion of respondents stated that they would be likely or very likely to choose to share a ride with a fellow passenger (not in their travel group) to get to the airport: 25% to share their own private car (if owned), 34% to share someone else's car, and 37% to share a taxi. Reasons put forward by respondents against sharing included unwillingness to share with a person not known to them (often due to personal safety) and a lack of convenience.

In conjunction with the various attitudinal data, this information was entered into a cluster analysis in order to identify a set of homogenous market segments of respondents, based on their surface access behavior, attitudes, situational characteristics and socio-demographic information. By establishing segments of passengers who share similar attitudes and characteristics, future policies can be targeted specifically to the groups where they are likely to stand the greatest chance of success.

Eight distinct groups were identified. These related to six groups who claimed to have regular access to a car in the UK, and two groups without car access. Each group is given a name based on the general attitudes and characteristics of the group. The two 'non-car access' groups were identified a priori, as access to a car inevitably heavily outweighs attitudinal or situational considerations when deciding how to travel to the airport. The eight groups are listed in Table 1, together with a very brief summary of each group's general outlook.

TABLE 1: A summary of the cluster profiles.

Cluster	Car access	Description
Devoted drivers (21.2%)	Yes	Very positive attitude towards car use, feel social pressure to do so. Negative view of public transport.
Remorseless motorists (17.6%)	Yes	Do not consider car access to airports to be a problem, and do not feel guilty about using their car.
Public transport avoiders (12.9%)	Yes	Not a particularly favourable attitude towards car use, but a very negative attitude towards public transport.
Frustrated drivers (11.1%)	Yes	Aware of the impacts of car use, but feel that using public transport is too difficult for them and would not make much of a difference to the overall problem.
Drop-offs (10.9%)	Yes	Very positive attitude towards being dropped-off at the airport, perceive large barriers to using public transport.
Conscientious 'greens' (4.6%)	Yes	Keenly aware of the negative effects of car use, have a positive attitude towards public transport and feel under social pressure to use it.
Riders of necessity (16.4%)	No	Neither a positive or negative attitude to public transport nor a relatively weak intention to use it in the future. More positive attitude towards taxi use.
Car-less crusaders (5.3%)	No	Very strong positive attitude towards public transport, perceive few barriers to using it, and feel that their own actions can make a difference.

The results of the cluster analysis revealed a wide range of attitudes and perceptions in the survey. As expected, the groups representing car users and people being dropped-off at the airport represented the largest share of the sample. From a policy perspective, it is important to target policies at groups where they are likely to stand the greatest chance of success. For example, policies to increase public transport use may be more effective when targeted at the 'Frustrated drivers' group than the 'Devoted drivers' group. The former are aware of the problem of car access to airports, but perceive large barriers to using public transport. In comparison, the 'Devoted drivers' have a very positive attitude towards using their car coupled with a negative view of public transport. A possible focus of policy in this case, for example, could be to try and reduce the perceived barriers to using public transport by the 'Frustrated drivers' group.

This also highlights an important point about travel behavior and modal choice, namely that people may act in the same way but for different reasons. For example, the 'Devoted drivers' and 'Public transport avoiders' groups both exhibit high private car use, but their reasons for this choice appear to be different. While the former have a positive attitude towards car use and value the comfort and convenience it provides, the latter appear to choose to use their cars because they actively dislike public transport, and want to avoid using it. This has important implications when formulating policy.

While significant attention is paid to increasing public transport use, it is equally important that strategies are implemented to ensure that there is not a mode shift in the 'wrong' direction. The 'Riders of necessity' group, for example, exhibit a higher than average use of public transport. It would seem that this not because of their positive view of public transport, but a result of their lack of access to a car. Indeed, attitudes to taxi and drop-off are more favorable than public transport use for this group. From an environmental perspective, it is important that there is not a significant shift towards taxi or drop-off use in this group, who represent the third largest segment in the analysis.

The two non-car owning clusters ('Riders of necessity' and 'Car-less crusaders') present an opportunity for change through technological innovation precisely because they do not have access to a car (i.e. a structural opportunity), whereas for the six car access clusters the opportunity relates more to altering their behavior (i.e. attitudinal opportunity). It can be assumed that the groups with the lowest propensity to use other modes

('Devoted drivers' and 'Remorseless motorists'), are also least likely to consider the use of alternative technologies. They are the clusters most likely to have established fixed patterns and routines with little consideration for other modes. That said, technology is arguably most important for the clusters dominated by drop-off / pick-up, and to a lesser extent car users, given that these trips are the least environmentally sustainable and the technologies (telepresence especially) are specifically aimed at reducing the use of these modes. There are three groups, 'Public transport avoiders', 'Frustrated drivers' and 'Drop-offs', that could be persuaded to use either public transport or share rides to reach airports. These groups, if facilitated with better routes and services such as RFID's on their luggage, could have confidence increased in public transport and hence take the next step and use it. Table 2 shows the three technological innovations and the clusters most appropriate for their application.

TABLE 2: A summary of the clusters appropriate for each of the three technological innovations.

Cluster	Technological innovation		
	Telepresence system	Public transport	Sharing rides
Public transport avoiders	*	*	*
Frustrated drivers	*	*	*
Drop-offs	*	*	*
Conscientious 'greens'	*		
Riders of necessity	*		
Car-less crusaders	*		

*Note: * = The technological innovation could be appropriate for this cluster.*

7.4 DISCUSSION AND CONCLUSIONS

Listing the surface access transport modes according to their environmental impact highlights the importance of the shift from motor car based trips to public transport. Initial investigation, incorporating the surface access transport mode 'pyramid' classification, has also shown the importance of distinguishing between 'double' trips (drop-off/pick-up and taxi) verses 'single' trips (car/park) to and from the airport. Using this framework to examine the sustainable development performance of surface access trips, the contribution of technology and behavioral intervention measures has been evaluated.

Although not the only solution, or a quick fix to the surface access problem, technology could have a role in reducing the carbon emissions generated from airport trips, particularly when the focus is on airports in the year 2020. Telepresence represents a technology that could reduce the level of drop-off / pick-up trips to the airport, as market presence increases and the costs of installation and usage decrease. Perhaps the use of technologies to encourage car share could also reduce drop-off / pick-up journeys, the most carbon emission generating method of travelling to and from airports. The Manchester airport survey has shown that a small but sizeable proportion of passengers would be willing to car share. A separate technological solution, the use of RFID tagging, could help to overcome the difficulty of taking luggage on public transport to access airports.

The behavioral element to the surface access problem has been examined using psychological-based theories and models. The link between intention and use of public transport has been confirmed; segmentation could be utilized to target particular population groups with the greatest propensity to use public transport, such as the 'Frustrated drivers' within the Manchester airport survey. The clusters generated can be linked to likelihood of using the technological innovations. In terms of marketing messages, it would not be effective to encourage individuals to use public transport for journeys to and from the airport using arguments that relate to moral feelings of guilt or efficacy. Although much of the focus here has been on the modal shift from the motor car to public transport, it has to be acknowledged that passengers can move in the opposite direction (from public transport to the motor car), and airports need to be aware of this.

The results of this study need to be placed in a wider context. Surface access transport journeys are dominated by motor car trips. Therefore, sustainable development gains are likely to be small in scale, and any improvements to the sustainable transport modal shares are likely to be only a few percentage points. These insights also need to be placed in an airport-related context. From the initial review it is evident that airports are not in control of all associated surface access operations, with a range of stakeholders with vested interests (e.g. public transport operators and third party tenants at the airport). In addition, as airports differ greatly in terms size and function, some of the solutions suggested here may not work in certain situations. For example, the technological solutions presented in this paper require an airport terminal of a sufficient size and passenger volume in order to be financially viable, and so may only work in the larger airports.

The work presented here will be developed to incorporate airport trips for employees and associated policy abatement methods such as car sharing schemes, working at home and incentives for using public transport. The research has also demonstrated (and as highlighted in [8]) that there is a further requirement to improve data collection at airports, such as the monitoring of emission targets in order to generate a clearer evidence base for emission improvements.

Much of the emphasis within this paper has been on environmental improvements (e.g. through carbon emission reductions), and as such the findings should also be placed in an economic context. Given that airports tend to be financially vulnerable and that there are current recessionary pressures, many of the proposed solutions within the paper need to be set in a medium- to long-term timeframe, say for the year 2020 and beyond.

REFERENCES

1. Civil Aviation Authority. UK Airport Statistics 2011; Civil Aviation Authority: London, UK, 2012. Available online: http://www.caa.co.uk/default.aspx?catid=80&pagetype=88&sglid=3&fld=2011Annual (accessed on 19 March 2013).
2. Department for Environment, Food and Rural Affairs. Securing the Future - UK Government Sustainable Development Strategy; HMSO: London, UK, 2005. Available online: http://archive.defra.gov.uk/sustainable/government/publications/uk-strategy/ (accessed on 19 March 2013).
3. Department for Transport. Towards a Sustainable Transport System. Supporting Economic Growth in a Low Carbon World; HMSO: London, UK, 2007. Available online: http://www.dft.gov.uk/about/strategy/transportstrategy/pdfsustaintranssystem.pdf (accessed on 19 March 2013).
4. Intergovernmental Panel on Climate Change-IPCC. Aviation and the Global Atmosphere. Key Report, 1999. Available online: http://www.grida.no/climate/ipcc/aviation/index.htm (accessed on 19 March 2013).

5.   Department for Environment, Food and Rural Affairs-DEPFRA, Climate Change Bill; HMSO: London, UK, 2008.

6.   Anderson, K.; Bows, A.; Footitt, A. Aviation in a Low Carbon EU. A Research Report by the Tyndall Centre; University of Manchester/Friends of the Earth: Manchester, UK, 2007; Available online: http://www.foe.co.uk/resource/reports/aviation_tyndall_07_main.pdf(accessed on 19 March 2013).

7.   Department for Transport. Aviation and Global Warming; Department for Transport: London, UK, 2004. Available online: http://www.dft.gov.uk/about/strategy/whitepapers/air/docs/aviationglobalwarmingreport.pdf (accessed on 19 March 2013).

8.   Humphreys, I.; Ison, S. Changing airport employee travel behaviour: The role of airport surface access strategies. Transp. Policy 2005, 1, 1–9.

9.   Vespermann, J.; Wald, A. Intermodal integration in air transportation: status quo, motives and future developments. J. Transp. Geogr. 2011, 19, 1187–1197.

10.   Koster, P.; Kroes, E.; Verhoef, E. Travel time variability and airport accessibility. Transport. Res. B 2011, 10, 1545–1559.

11.   Hess, S.; Polak, J.W. Exploring the potential for cross-nesting structures in airport-choice analysis: A case study of the Greater London Area. Transport. Res. E 2006, 42, 63–81.

12.   Budd, T.M.J.; Ison, S.G.; Ryley, T.J. Airport surface access management: Issues and policies. Journal of Airport Management 2011, 1, 80–97.

13.   Budd, T.M.J.; Ison, S.G.; Ryley, T.J. Airport surface access in the UK: A management perspective. RTBM 2011, 1, 100–117.

14.   Manchester Airport. Ground Transport Plan. Part of the Manchester Airport Master Plan to 2030; Manchester Airport: Manchester, UK, 2007. Available online: http://www.manchesterairport.co.uk/manweb.nsf/AttachmentsByTitle/TransportStrategy/$FILE/Grndtrans-screen.pdf (accessed on 19 March 2013).

15.   Civil Aviation Authority. CAA Passenger Survey Report; Civil Aviation Authority: London, UK, 2009. Available online: http://www.caa.co.uk/default.aspx?catid=81&pagetype=90&pageid=7640 (accessed on 19 March 2013).

16.   Miyoshi, C.; Mason, K. The damage cost of carbon dioxide emissions produced by passengers on airport surface access: The case of Manchester Airport. J. Transp. Geogr. 2013, 28, 137–143.

17.   Future Travel Experience. Offsite Agents to Improve Passenger Service Levels, 2012. Available online: http://www.futuretravelexperience.com/2012/05/offsite-agents-to-improve-passenger-service-levels/ (accessed on 19 March 2013).

18.   Ajzen, I. The theory of planned behavior. Organ. Behav. Human Dec. 1991, 50, 179–211.

19.   Schwartz, S.H. Normative Ifluences on Altruism. In Advances in Experimental Social Psychology; Berkowitz, L., Ed.; Academic Press: New York, NY, USA, 1977; pp. 221–279.

20.   Bamberg, S.; Fujii, S.; Friman, M.; Gärling, T. Behaviour theory and soft transport policy options. Transp. Policy 2011, 1, 228–235.

21.   Abrahamse, W.; Steg, L.; Gifford, R.; Vlek, C. Factors influencing car use for commuting and the intention to reduce it: A question of self-interest or morality? Transport. Res. F 2009, 4, 317–324.

22. Bamberg, S.; Hunecke, M.; Blöbaum, A. Social context, personal norms and the use of public transportation: Two field studies. J. Environ. Psychol. 2007, 3, 190–203.

23. Pels, E.; Nijkamp, P.; Rietveld, P. Access to and competition between airports: A case study for the San Francisco Bay Area. Transport. Res. A 2003, 1, 71–83.

24. Dresner, M. Leisure verses business passengers: Similarities, differences, and implications. J. Air Transp. Manag. 2006, 1, 28–32.

25. Leigh Fisher Associates; Coogan, M.A.; MarketSense Consulting. Improving Public Transportation Access to Large Airports; TCRP (Transit Cooperative Research Program) Report 62; Transportation Research Board of the National Academies: Washington, DC, USA, 2000.

26. Brilha, N.M. Airport Requirements for Leisure Travellers. In Aviation and Tourism: Implications for Leisure Travel; Graham, A., Papatheodorou, A., Forsyth, P., Eds.; Ashgate: London, UK, 2008; pp. 167–177.

PART IV

FOSSIL FUEL ALTERNATIVES

CHAPTER 8

Diesel Internal Combustion Engine Emissions Measurements for Methanol-Based and Ethanol-Based Biodiesel Blends

CHARALAMBOS A. CHASOS, GEORGE N. KARAGIORGIS, AND CHRIS N. CHRISTODOULOU

8.1 INTRODUCTION

There is recent interest for the utilisation of renewable and alternative fuels by the European Union (EU), which is regulated by Directive 2009/30/EC. For the usage of biodiesel blends, a lower limit of 7% by volume biodiesel fuel blend in diesel fuel is currently imposed. The specifications of biodiesel fuels which can be used for blending diesel fuel are defined by the European standard EN 14214 [1]. The physical properties of biodiesel affect the diesel blends, and the range of density, viscosity, and flash point of biodiesel are specified in EN 14214. However, other physical properties

Diesel Internal Combustion Engine Emissions Measurements for Methanol-Based and Ethanol-Based Biodiesel Blends. © *Chasos CA, Karagiorgis GN, and Christodoulou CN.* Conference Papers in Energy **2013** *(2013), http://dx.doi.org/10.1155/2013/162312. Licensed under Creative Commons Attribution 3.0 Unported License, http://creativecommons.org/licenses/by/3.0/.*

of diesel blends including the surface tension coefficient, the fuel vapour pressure, the boiling point, and the latent heat of evaporation also affect the injected fuel spray characteristics, the resulting air/fuel mixing, combustion, and the emissions of diesel internal combustion engines (ICE), as well as the engine overall performance.

Biodiesel is mainly produced from oilseed crops and other raw materials [2]. Biodiesel fuels can be produced from rapeseed via cold pressing/extraction and transesterification, known as fatty acid methyl ester (FAME) [3], and are known as first generation biofuels. For second generation biodiesel fuels known as hydrotreated biodiesel, hydrotreatment technologies are used for vegetable oils and animal fat materials [3].

Biodiesel fuel type and fuel physical properties, as well as the resulting physical properties of different blends of biodiesel in diesel fuel, are considered very important but were studied in a limited number of previous experimental and computational investigations. The biodiesel physical properties including density, dynamic viscosity, and surface tension coefficient have been investigated by [4] among others. At atmospheric conditions, the density of biodiesel compared to pure diesel fuel is around 5 to 10% higher, and the viscosity of biodiesel given at 40°C is higher by almost factor of two than the viscosity of pure diesel. Limited data was published for surface tension coefficient, which ranges from 0.025 to 0.03 N/m for biodiesel, while for diesel ranges approximately from 0.02 to 0.025 N/m. However, the different raw materials as well as the different production processes for biodiesel result in different physical properties of the produced biodiesel. The present work aims to examine the effect of different raw materials, namely, methanol and ethanol, used for the production of biodiesel in order to distinguish their effects on the exhausted emissions and engine performance.

Published experimental and computational studies dealt with diesel injector internal flow, cavitation phenomena and the resulting sprays for biodiesel fuel. The resulting sprays from diesel fuel and biodiesel blends were examined experimentally in [5, 6] among others. It was found that the spray cone angle decreases and spray penetration increases with increasing the blending percentage of biodiesel. Computational spray studies were performed in [7] among others, and it was found that spray cone angle decreases and spray penetration increases when the blending percentage of biodiesel is increasing.

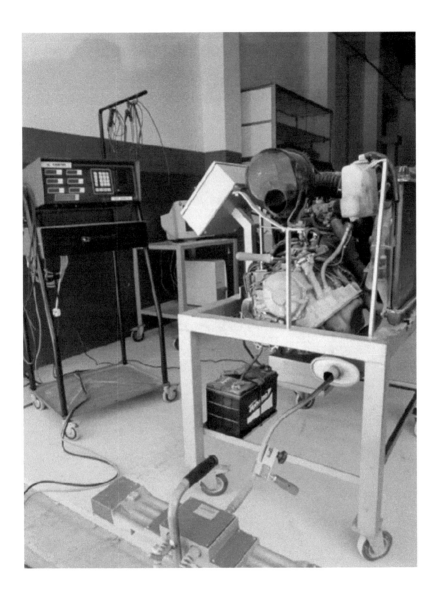

FIGURE 1: Diesel ICE on a frame, used for biodiesel blends testing.

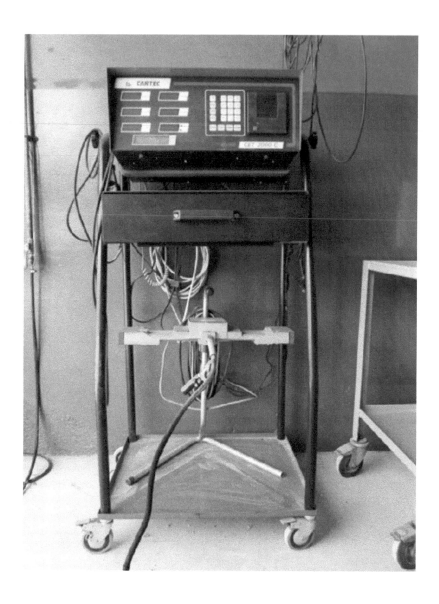

FIGURE 2: Gas analyser with exhaust emissions probe.

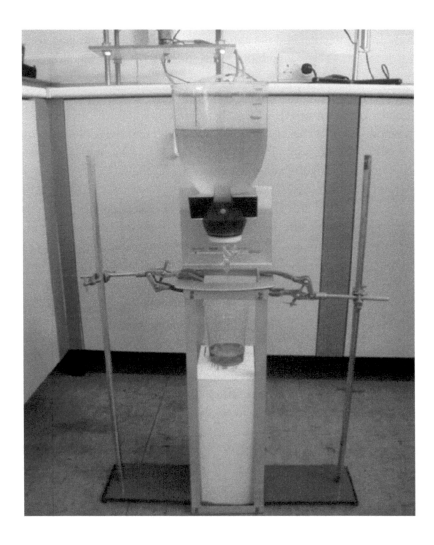

FIGURE 3: Photograph of the experimental apparatus for the production of methanol-based biodiesel, separation of glycerine and other by-products, and collection of biodiesel.

In published experiments the effects of different types of biodiesel blends on combustion and the exhaust emissions and performance of diesel ICE (see [8–11]) were studied. Anand et al. [8] used a nondispersive infrared analyser (NDIR) (AVL DiGas 444), gas analyser, and smoke meter (AVL437) and measured the emissions of blends of waste cooking oil methyl ester from B10 (10% by volume biodiesel in diesel) to B80 in a single-cylinder diesel engine. It was found that, with biofuel blends, the specific fuel consumption increased, the CO, CO_2, and HC emissions decreased, the smoke opacity decreased, while the NO emission slightly increased [8]. Guido et al. [9] employed a smoke meter (AVL415S) and studied biodiesel blends B20, B50, and B100 effects on the emissions of a General Motors 2-litre four-cylinder diesel engine. It was also found that the emitted smoke is reduced with increasing percentage of biodiesel blends [9]. Nabi and Hustad [10] employed a gravimetric method and examined the 20% in volume jatropha biodiesel blend in diesel in a six-cylinder turbocharged direct injection engine, and found that smoke decreased with the use of blended fuel. Zhang et al. [11] used a portable exhaust gas analyzer (FGA-4100) along with a smoke opacity meter (AVL 439) and investigated the combustion and emission characteristics of soybean methyl ester biodiesel blends in a single-cylinder direct injection engine at different loads and constant speed. It was found that smoke opacity decreases with higher percentage of biodiesel blends [11]. In a recent investigation [12], the effects of increasing blending percentage of biodiesel on the emitted smoke from the engine utilising gas analyser described in the present study were examined. It was found that the amount of emitted smoke decreases with increasing biodiesel blends, which agrees with the other published investigations.

However, further experimental investigations accompanied with computational studies should be carried out for better understanding of the quality of biodiesel and the various effects of biodiesel on emissions and performance of ICE. Particularly, the effect of different types of biodiesel fuels is required to be assessed.

The present study examines, experimentally, vehicle diesel internal combustion engines operating with blends of two different types of biodiesel fuels, namely, methanol-based and ethanol-based denoted MB and

EB, respectively, in diesel fuel. The effects of different types of biodiesels blends on ICE emissions are identified. From the experimental results and discussion, conclusions and suggestions are provided regarding the adaptation of biodiesel in ICE. First, the experimental setup used for the measurements of the emissions is presented, including details for the production of the two different types of biodiesel fuels which were used for the blending of pure diesel fuel at increasing percentages. Then, the experimental results are discussed, followed by the conclusions of the present study.

TABLE 1: Technical specifications of the diesel ICE.

Description	Details
Engine code	4D68
Engine type	Inline 4 cylinders, turbocharged diesel
Displacement (cm3)	1998
Bore (mm)	82.7
Stroke (mm)	93
Compression ratio	22.4

8.2 EXPERIMENTAL SETUP

The experimental setup used for the measurements includes the diesel internal combustion engine where the test fuels were examined and the exhaust emissions were measured, the exhaust gas analyser which was employed for the measurements, the laboratory production of the two types of biodiesel, and the preparation of blends at various percentages.

8.2.1 DIESEL INTERNAL COMBUSTION ENGINE

The diesel ICE used for the measurements of emissions for pure diesel fuel (Eurodiesel) and different blends of biodiesel in Eurodiesel is shown in Figure 1. The diesel ICE is manufactured by Mitsubishi and is mounted on a special antishock frame. It is a four-cylinder engine with four valves per cylinder (2 inlet and 2 exhaust valves) with direct injection fuel system equipped with a turbocharger. Technical details of the engine are tabulated in Table 1.

8.2.2 GAS ANALYSER AND EXPERIMENTAL PROCEDURE

The gas analyser which was used for the exhaust emissions measurements from the diesel ICE is shown in Figure 2. The test fuels of various blends of the two different types of biodiesel fuels in diesel fuel were utilised for fuelling the ICE. The gas analyser model is CARTEC CET 2000 [13], and it has various sensors and probes for measurements of exhaust emissions and monitoring of engine speeds in revolutions per minute (rpm).

The gas analyser can measure gasoline ICE emissions and diesel ICE emissions [13]. The absorption method utilising turbidity meter (opacimeter) is used for the diesel engine emissions ([13, 14]). For diesel ICE, the measured emitted smoke is given by the coefficient of light obscuration (CLO) which is analogous to the mass concentration of unburned carbon particles contained in the exhaust gases. The corresponding values of smoke as mass concentration to the CLO are given by the manufacturer in tables. The measurement of CLO as function of engine speed and engine temperature is used in the presentation of the experimental results.

8.2.3 METHANOL-BASED BIODIESEL PRODUCTION AND PREPARATION OF BLENDS WITH PURE DIESEL

The methanol-based biodiesel was produced in the Materials Process Laboratory of Frederick University and was used for the preparation of blends of biodiesel in conventional diesel fuel. The biodiesel production and the preparation of blends with conventional diesel are described below.

The raw materials included the corn oil, methanol, and sodium hydroxide (NaOH) ([15, 16]). For every 1 liter of corn oil, 250 mL of methanol and 3.5 g of NaOH were used. The production process involves heating of the corn oil at around 50–60°C and adding of methanol with dissolved

NaOH by constant stirring for a period of about 20 minutes. The mixture was left overnight for liquid phase separation. Biodiesel is the lighter liquid phase, whereas the heavier liquid is the biowaste containing glycerine and other by-products (mainly, NaOH and moisture). The biowaste was emptied by draining, while biodiesel was left in the container as shown in Figure 3.

The methanol-based biodiesel blends with conventional diesel were prepared in the ICE laboratory with mixing of the produced methanol-based biodiesel in various percentages with conventional diesel fuel obtained from local market [17], traded as "Eurodiesel." The test fuels were pure Eurodiesel, denoted "diesel," and 25, 50, and 75% percentage of methanol-based biodiesel volume in the total volume of blended diesel, denoted "MB25, MB50" and "MB75", respectively.

8.2.4 ETHANOL-BASED BIODIESEL PRODUCTION AND PREPARATION OF BLENDS WITH PURE DIESEL

The ethanol-based biodiesel was also produced in the Material Process Laboratory of Frederick University and was used for the preparation of blends of ethanol-based biodiesel in conventional diesel fuel. For the ethanol-based biodiesel production, ethanol was used instead of methanol. For every one liter of corn-oil, 350 mL of ethanol was used, while the same amount of NaOH was used. The processing of ethanol-based biodiesel resulted in increased amount of produced biowaste as shown in Figure 4.

The ethanol-based biodiesel biowaste depicted in Figure 4 was thicker and different in colour than the biowaste of the methanol-based biodiesel shown in Figure 3. Thus, the production process of ethanol-based biodiesel required washing of biodiesel with water and infiltration in order to improve the production and quality of the produced biodiesel.

The ethanol-based biodiesel blends with conventional diesel were prepared in the ICE laboratory with mixing of the produced ethanol-based biodiesel in various percentages with Eurodiesel. The test fuels were pure Eurodiesel, denoted by "diesel," and 10 and 25% percentage of ethanol-based biodiesel volume in the total volume of blended diesel, denoted by "EB10," and "EB25," respectively.

FIGURE 4: Photograph of the experimental apparatus for the production of ethanol-based biodiesel, separation of biowaste and other by-products, and collection of biodiesel.

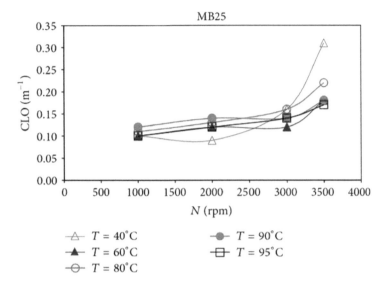

FIGURE 5: Smoke emissions measurements as function of engine speed at increasing engine temperatures for pure diesel.

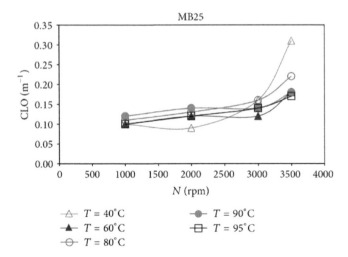

FIGURE 6: Smoke emissions measurements as function of engine speed at increasing engine temperature for MB25 test fuel.

8.3 EXPERIMENTAL RESULTS AND DISCUSSION

The measurements of exhaust emissions were carried out for increasing engine speeds from idle to full throttle. The measurements started from engine at atmospheric temperature and were recorded for increasing engine temperature until fully warm-up. For the measurement of the exhaust emissions which were produced from the test fuels, the engine speed, the lubricant oil temperature (corresponding to the engine warm-up evolution), and the emitted smoke expressed with the CLO in (m^{-1}) were recorded for lubricant oil temperature approximately around 40, 60, 80, 90, and 95°C, at engine speeds of 1000, 2000, 3000, and 3500 revolutions per minute (rpm).

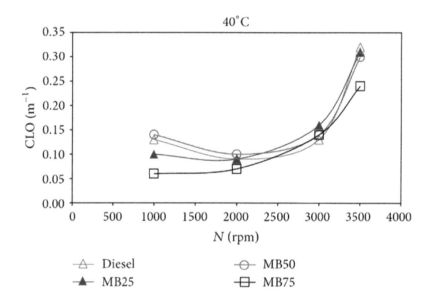

FIGURE 7: Smoke emissions measurements at cold engine temperature as function of engine speed for increasing blending percentage of methanol-based biodiesel in diesel.

The experimental results of the emissions of pure diesel blends are firstly described, then the emissions of methanol-based biodiesel blends are presented, followed by the experimental results of the emissions of the ethanol-based biodiesel blends. Finally, the emissions of methanol-based and ethanol-based biodiesel blends are compared against the emissions of pure biodiesel and discussed.

8.3.1 EMISSIONS RESULTS OF PURE BIODIESEL

The measurements of emissions from the diesel ICE for pure diesel at engine temperature approximately 40, 60, 80, 90, and 95°C at increasing engine speeds are presented in Figure 5. It can be seen that the emitted smoke decreases slightly with a minimum value at 2000 rpm, and increases considerably when the engine speed is higher than 3000 rpm for all the engine temperatures.

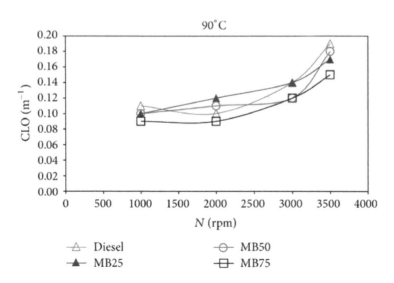

FIGURE 8: Smoke emissions measurements at hot engine temperature as function of engine speed for increasing blending percentage of methanol-based biodiesel in diesel.

As it is shown in Figure 5, the maximum amount of emitted smoke occurs for cold engine operation at maximum engine speed, which reveals that there is no sufficient time for the air/fuel mixture to be prepared and combustion to be completed.

8.3.2 EMISSIONS RESULTS OF METHANOL-BASED BIODIESEL BLENDS

The measurements with methanol-based biodiesel blend MB25 are included in Figure 6. For all engine temperatures, the emitted smoke amount slightly increases in a linear trend with the increasing engine speed until 3000 rpm. Thereafter, it can also be seen that at the peak speed, the rate of increase of the smoke amount increases and that the maximum quantity of smoke is emitted for cold engine operation. The trends that are observed are similar to the trends of pure diesel fuel, thus the engine operation according to the emissions does not deteriorate.

The measurements of smoke with methanol-based biodiesel blends for the test fuels MB25, MB50, and MB75 are compared against the emitted smoke from pure diesel fuel in Figure 7. Figure 7 shows the effect of engine speed for cold engine operation for the different test fuels. It can be observed that smoke decreases slightly from 1000 to 2000 rpm for all the test fuels except for MB75, and then the emitted smoke slightly increases until 3000 rpm. For engine speeds greater than 3000 rpm, the emitted smoke increases substantially. The differences between emitted amount of smoke for diesel fuel, MB25, and MB50 are small, while the lower amount of emitted smoke takes place for the MB75, which shows that increasing the blending percentage of methanol-based biodiesel reduces the amount of smoke for all engine speeds. Therefore, for cold engine operation, the utilization of increased percentage of biodiesel is considered beneficial for the engine operation.

The measurements of smoke with methanol-based biodiesel blends for the test fuels diesel, MB25, MB50, and MB75 for hot engine operation for increasing engine speeds are compared in Figure 8. Figure 8 shows that smoke slightly increases from 1000 to 3000 rpm for all the test fuels. For engine speeds higher than 3000 rpm, the emitted smoke rate of increase is higher than the rate for speeds lower than 3000 rpm. The differences between emitted amount of smoke for all test fuels are negligible for the range of speeds from 1000 rpm to 3000 rpm, and for higher engine speeds, the emitted smoke decreases by around 20%, when the blending percentage increases from 0 to 75%.

Thus, the trend observed in Figure 8 is similar regarding the levels of emitted amount of smoke at high engine speeds, and increasing the blending percentage of methanol-based biodiesel it can improve the engine performance. It is obvious that, for hot engine operation, using increased percentage of biodiesel in the blend improves air/fuel mixing and combustion quality especially at high engine speeds.

Overall, from Figures 5, 6, 7, and 8, it can be observed that the emitted smoke trends are rather similar for pure diesel, MB25, MB50, and MB75. However, the highest amount of smoke is emitted when the engine is cold and running at maximum speed with pure diesel fuel, and the lowest amount of smoke is exhausted when the engine is hot and operating at low engine speed when the fuel blend has the maximum blending percentage of methanol-based biodiesel.

8.3.3 EMISSIONS RESULTS OF ETHANOL-BASED BIODIESEL BLENDS

The measurements with ethanol-based biodiesel blend EB25 are included in Figure 9. For all engine temperatures, the emitted smoke amount is almost the same for engine speeds up to 3000 rpm and for engine temperatures higher than 60°C. For 3000 rpm, the smoke amount increases dramatically, and the maximum quantity of smoke is emitted for cold engine operation. This reveals that the engine operation worsens when the engine is cold and operating with ethanol-based biodiesel blends even at low percentages. This can be explained by the different physical properties of the ethanol-based biodiesel which result in poor air/fuel mixing, evaporation, and incomplete combustion. Test measurements were performed with test fuels at higher blending percentage than 25% of ethanol-based biodiesel blends, which resulted in unstable engine operation, when vibrations and heavy noise occurred.

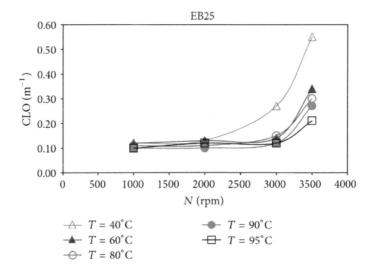

FIGURE 9: Smoke emissions measurements as function of engine speed at increasing engine temperature for EB25 test fuel.

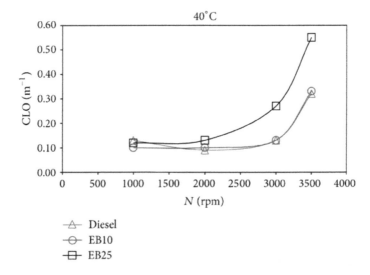

FIGURE 10: Smoke emissions measurements at cold engine temperature as function of engine speed for increasing blending percentage of ethanol-based biodiesel in diesel.

The measurements of smoke with ethanol-based biodiesel blends for the test fuels EB10 and EB25 are compared against the emitted smoke from pure diesel fuel in Figure 10. Figure 10 shows the effect of engine speed for cold engine operation for three test fuels at increasing blending percentage of ethanol-based biodiesel. It can be observed that the amount of smoke is almost constant for 1000 to 2000 rpm for all the test fuels. However, for higher engine speeds, the emitted smoke increases dramatically for the EB25. The differences between emitted amount of smoke for diesel and EB10 test fuels are negligible, which suggests that ethanol-based biodiesel can be used without inhibiting engine's performance. However, it is prohibited to use ethanol-based biodiesel at increasing blending percentages in the blends since the engine operation dramatically deteriorates.

The measurements of smoke with ethanol-based biodiesel blends for the test fuels diesel, EB10 and, EB25 for hot engine operation at increasing engine speeds are presented in Figure 11. Figure 11 shows that smoke slightly increases in the speed range of 1000 to 3000 rpm for all the test fuels. For engine speeds higher than 3000 rpm, the emitted smoke amount increases dramatically for EB10 and EB25, while for diesel fuel the smoke increases by around 20%. The differences between the emitted amount of smoke for all test fuels are negligible for the range of speeds from 1000 rpm to 3000 rpm, and for higher engine speeds, the emitted smoke increases when blending percentage is increasing.

The trend observed in Figure 11 regarding the increase of smoke with increasing engine speed for hot engine operation is similar to the trend shown in Figure 10, regarding the levels of emitted amount of smoke. However, at high engine speeds, increasing the blending percentage of ethanol-based biodiesel increases dramatically the amount of emitted smoke, and this is more evident during hot engine operation. Thus, for both cold and hot engine operation, increasing the blending percentage of ethanol-based biodiesel in the blend affects the air/fuel mixing and the combustion quality especially at high engine speeds.

Overall, from Figures 9, 10, and 11, it is found that EB25 and fuels with higher blending percentages should not be used for blending diesel fuel.

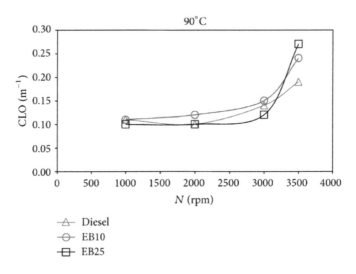

FIGURE 11: Smoke emissions measurements at hot engine temperature as function of engine speed for increasing blending percentage of ethanol-based biodiesel in diesel.

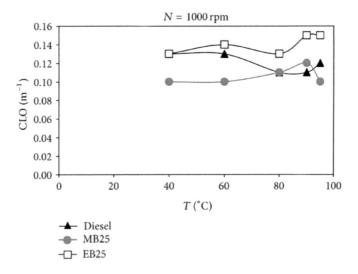

FIGURE 12: Smoke emissions measurements at low engine speed (1000 rpm) as function of engine temperature for pure diesel, MB25, and EB25 test fuels.

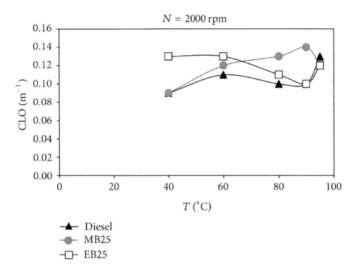

FIGURE 13: Smoke emissions measurements at medium engine speed (2000 rpm) as function of engine temperature for pure diesel, MB25, and EB25 test fuels.

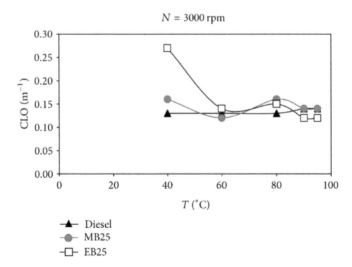

FIGURE 14: Smoke emissions measurements at high engine speed (3000 rpm) as function of engine temperature for pure diesel, MB25, and EB25 test fuels.

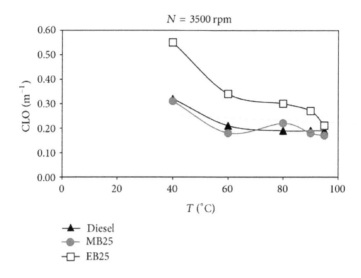

FIGURE 15: Smoke emissions measurements at high engine speed (3500 rpm) as function of engine temperature for pure diesel, MB25, and EB25 test fuels.

8.3.4 EMISSIONS COMPARISONS OF METHANOL-BASED AND ETHANOL-BASED BIODIESEL BLENDS

The comparisons of the measurements of emitted smoke from the diesel ICE for pure diesel, MB25, and EB25 test fuels at engine temperature approximately 40, 60, 80, 90, and 95°C at low, medium, and high engine speeds are presented below.

In Figure 12, the low engine speed results are included, and it can be seen that the amount of smoke fluctuates and the levels are low for all test fuels. However, the lowest amount of smoke is produced with methanol-based biodiesel and the maximum amount of smoke is emitted with ethanol-based biodiesel blend. It can be seen that using methanol-based biodiesel reduces the amount of smoke compared to diesel and EB25 at low engine temperatures, whereas using ethanol-based biodiesel blend increases the amount of smoke compared to diesel and methanol-based biodiesel blend at hot engine conditions.

In Figure 13, the emitted amount of smoke is compared for pure diesel, MB25, and EB25 test fuels at medium engine speed. The differences are small, but it is evident that EB25 results in higher amount of smoke for cold engine operation. However, at hot engine conditions the level of emitted smoke is almost the same for all test fuels.

Increasing the engine speed to high levels results in similar trends for the emitted smoke with diesel and MB25 as shown in both Figures 14 and 15. The first trend is that, for diesel and MB25 test fuels, the emitted smoke slightly decreases when the engine temperature is increasing. The second trend is that the emitted amount of smoke for EB25 decreases when the engine temperature is increasing, which reveals that the air/fuel mixing, evaporation, and combustion quality are improved. This can be explained by the improved physical properties of the ethanol-based bio-fuel blend when heating takes place in the cylinder of the diesel ICE.

8.4 CONCLUSIONS AND RECOMMENDATIONS

For increasing blending percentage of methanol-based biodiesel blends, the highest amount of smoke is emitted when the engine is cold and running at maximum speed with pure diesel fuel, and the lowest amount of smoke is emitted when the engine is hot and operating at low engine speed when the fuel blend has the maximum blending percentage of methanol-based biodiesel. The adaptation of methanol-based blends is recommended because they reduce smoke.

At higher engine speeds the emitted smoke increases when blending percentage is increasing for ethanol-based biodiesel blends. However, ethanol-based biodiesel blends at low blending percentage can be used, but should be further investigated. However, blends with higher blending percentages should not be used for blending diesel fuel. It is recommended to examine preheating of ethanol-based biodiesel blends at increasing blending percentages.

For both methanol-based biodiesel blends and ethanol-based biodiesel blends, further investigations on NOx emissions quality, as well as the effects of blended diesel on engine thermal efficiency and brake power, should be performed. Furthermore, other types of diesel engines should be used, including naturally aspirated engines, for investigations of effects of biodiesel blends on engine performance and emissions.

The physical properties of biodiesel fuels, including methanol-based biodiesel and ethanol-based biodiesel, and resulting blends should be examined experimentally and computationally in order to relate and quantify the effects of their physical properties on the diesel engine emissions and engine overall performance.

REFERENCES

1. Cyprus Organisation for Standardisation (CYS), CYS EN, 14214:2008+A1: automotive fuels: Fatty acid methyl esters (FAME) for diesel engines: requirements and test methods, 2009.
2. European Commission, Factsheet, Biofuels in the European Union: An Agricultural Perspective, Brussels, Belgium, 2006.
3. European Commission, "Biofuels in the European Union, a vision for 2030 and beyond," Final Report of the Biofuels Research Advisory Council, Brussels, Belgium, 2006, Directorate-General for Research, Sustainable Energy Systems.
4. Y. Ra, R. D. Reitz, J. Mc Farlane, and C. S. Daw, "Effects of fuel physical properties on diesel engine combustion using diesel and Bio-diesel fuels," in Proceedings of the Society of Automotive Engineers, SAE International Journal of Engines, Detroit, Mich, USA, April 2008.
5. B. S. Higgins, C. J. Mueller, and D. L. Siebers, "Measurements of fuel effects on liquid-phase penetration in DI sprays," in Proceedings of the Society of Automotive Engineers, SAE International Journal of Engines, Detroit, Mich, USA, March 1999.
6. G. Valentino, L. Allocca, S. Iannuzzi, and A. Montanaro, "Biodiesel/mineral diesel fuel mixtures: spray evolution and engine performance and emissions characterization," Energy, vol. 36, no. 6, pp. 3924–3932, 2011.
7. C. A. Chasos, C. N. Christodoulou, and G. N. Karagiorgis, "CFD simulations of multi-hole Diesel injector nozzle flow and sprays for various biodiesel blends," in Proceedings of 12th Triennial International Conference on Liquid Atomization and Spray Systems (ICLASS '12), Heidelberg, Germany, September 2012.
8. R. Anand, G. R. Kannan, S. Nagarajan, and S. Velmathi, "Performance emission and combustion characteristics of a diesel engine fueled with biodiesel produced from waste cooking oil," in Proceedings of the Society of Automotive Engineers, SAE International Journal of Engines, Detroit, Mich, USA, 2010.
9. C. Guido, C. Beatrice, S. Di Iorio et al., "Alternative Diesel fuels effects on combustion and emissions of an Euro 5 automotive Diesel engine," in Proceedings of the

Society of Automotive Engineers, SAE International Journal of Engines, Detroit, Mich, USA, April 2010.

10. M. N. Nabi and J. E. Hustad, "Effect of fuel oxygen on engine performance and exhaust emissions including ultrafine particle fueling with Diesel-oxygenate blends," in Proceedings of the Society of Automotive Engineers, SAE International Journal of Engines, San Diego, Calif, USA, October 2010.

11. X. Zhang, G. Gao, L. Li, Wu, Z. Hu Z, and J. Deng, "Characteristics of combustion and emissions in DI engine fueled with biodiesel blends from soybean oil," in Proceedingsof the Society of Automotive Engineers, SAE International Journal of Engines, Shanghai, China, June 2008.

12. C. A. Chasos, E. I. Ioannou, A. E. Kouroufexis, C. N. Christodoulou, P. M. Artemi, and G. N. Karagiorgis, "Biofuels production and testing in Internal Combustion Engines," in Proceedings of the 3rd International Conference on Renewable Energy Sources and Energy Efficiency, Nicosia, Cyprus, May 2011.

13. CARTEC Operation Manual, "Engine exhaust gas analysis CET 2000," Tech. Rep., CARTEC Richard Langlechner GmbH, Unterneukirchen, Germany.

14. Bosch Automotive Handbook, Bentley Publishers, 6th edition, 2004.

15. Medisell Co. Ltd, Laboratory Supplies. Nicosia, Cyprus, http://www.medisell.com.cy.

16. Ambrosia Oils (1976) Ltd. Larnaka, Cyprus, http://www.ambrosia.com.cy.

17. Hellenic Petroleum Cyprus Ltd, http://www.eko.com.cy.

CHAPTER 9

Electric Cars: Technical Characteristics and Environmental Impacts

ECKARD HELMERS AND PATRICK MARX

9.1 INTRODUCTION

On a worldwide scale, 26% of primary energy is consumed for transport purposes, and 23% of greenhouse gas emissions is energy-related. Street traffic represents a share of 74% in the transport sector worldwide (IPCC data from 2007, as summarized in [1]). The transport sector includes aircraft, ships, trains, and all types of street vehicles (e.g., trucks, buses, cars and two-wheelers). Automobiles play a particular role for three reasons: First, cars are dominating the street traffic in most countries. Second, car sales exhibit the greatest growth rates in the world. Third, there are alternative technologies for the drivetrain available unlike, e.g., for trucks. While small trucks may also be operated electrically within a limited range, big trucks are dependent on diesel fuel, which can be shifted to a mixture of 80% methane (either fossil or biogenic) in the future. Buses can also be driven electrically on limited distances; buses driven by compressed natural gas (methane) are routinely used. While fuel cell-driven buses are already on the streets, small trucks driven by fuel cells and H_2 are still concepts.

In Germany, for example, cars are responsible for 60% of all traffic-related CO_2 emissions (German Federal Environment ministry number for 2010, summarized in [1]). In the future, traffic is expected to grow enormously worldwide, particularly in developing Asian countries. The worldwide vehicle stock of 630 million may grow to one billion in 2030 (data from Shell 2007, reviewed by Angerer et al. [2]). Vehicle production is expected to grow from 63 to 100 million cars per year until 2030 [2]. In addition to the CO_2 emissions, modern internal combustion engine vehicles (ICEVs) still have dangerous toxic emissions. According to the World Health Organization (WHO) [3], air pollution is a major environmental risk for health and is estimated to cause approximately two million premature deaths worldwide per year. Since ozone, fine dust, NO_2, and SO_2 have been identified by WHO as being the most dangerous kinds which are mainly, or to a substantial extent, traffic-derived, traffic will be responsible for approximately half of that quantified costs in lives and health. Toxic ICEV emissions cause high health costs even in industrialized countries: Almost 25% of the European Union (EU)-25 population live less than 500 m from a road carrying more than three million vehicles per year. Consequently, almost four million years of life are lost each year due to high pollution levels (press release European Environmental Agency, 26 February 2007).

In order to meet future mobility needs, reduce climate as well as health relevant emissions, and phase out dependence on oil ('peak oil'), today's propulsion technologies have to be replaced by more efficient and environmentally friendly alternatives. On the transition to a sustainable society, particularly efficient mobility technologies are needed worldwide. Electric vehicles have been identified as being such a technology [4]. In parallel, a couple of countries (like Germany, Denmark, and Sweden) have decided to switch electricity production from fossil fuel to renewable sources, further improving sustainability of electric cars when compared with ICEV.

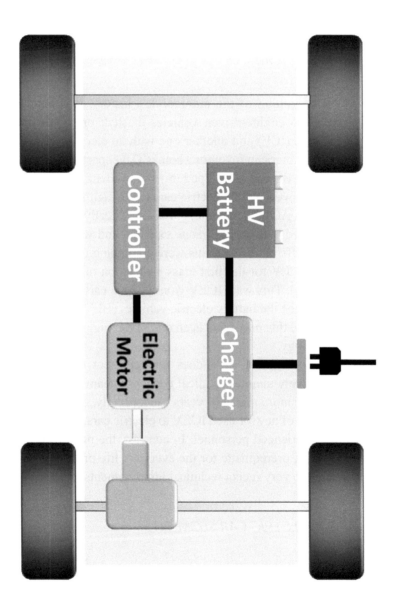

FIGURE 1: Important components of an electric car. (HV, high voltage).

9.2 TECHNOLOGY

9.2.1 CONCEPTS OF ELECTRIC CARS

9.2.1.1 HISTORY

At the beginning of the automobile's history, two main competing approaches to engine-driven vehicles existed: one with internal combustion engine (ICE) and another one with an electric drivetrain. Already in 1834, the American inventor Thomas Davenport built the first electric car. The first ICEV was developed in 1886 by Benz and Daimler in Germany. Around the year 1900, electric cars had a significant share of all engine-driven cars. At the same time, F. Porsche already invented a hybrid electric car equipped with an ICE range extender and wheel hub electric engines. The two different drive trains were competing until Henry Ford, in 1908, chose an ICEV for the first mass production of a car in history (summarized in [5]). This way, ICEV won the race early in the twentieth century and displaced the battery electric vehicles (BEV). From an environmental perspective, this may have been one of the biggest mistakes in the history of technology.

Concluding, the BEV does not represent recent 'high tech', but a comparatively simple technical concept, meanwhile available as a series product for more than 110 years. Accordingly, e-conversion, which is the conversion of new or used ICEV to electric cars, can easily be implemented by experienced personnel. In contrast, the modern lithium-ion battery technology, prerequisite for the everyday life practicability of most BEV, is related to very recent technical improvements.

9.2.1.2 ELECTRIC CAR SEGMENTS

Within the 1990s, electric cars were again offered as series products in California due to its Zero Emission Act (summarized in [5]). After the Zero Emission Act had been suspended, Partial Zero Emission cars were preferred by the Californian government, which prompted the carmaker

Toyota to develop the hybrid vehicle, combining electric and combustion engine. Energy efficiency improved drastically this way (see below); also, the idea of the electric car spread worldwide parallel to the success of the Toyota Prius. Since a full hybrid vehicle is able to drive electrically, it simply needs a plug and a bigger battery in order to be charged like a BEV. This way, the category of plug-in hybrid vehicles (PHEV) was created. Within the last 10 years, different drivetrain concepts based on electric motors have been developed and are soon going to enter mass production. All-electric drive and hybrid electric drive have to be differentiated. In contrast to the hybrid electric drive, in the all-electric car, an electric motor is the only energy converter. According to a UN definition from 2003, a hybrid electric drivetrain consists of at least two different energy converters (e.g., ICE and electric motor) contributing to the drive system and uses two different energy storages (e.g., fuel and battery) (see, e.g., UNEP [6]).

TABLE 1: Possible market segments of BEV, PHEV, and FCV (after Nemry and Brons[8], completed)

Size of car	BEV	PHEV	FCV
Small	Immediate candidate	Vehicle packaging problem and high price are obstacles.	
	Most useful according to practicability reasons related to battery size and costs		
Medium	Useful	Privileged segment. Long distance trips possible. However, H2 filling station infrastructure required for FCV.	
	Some models expected within forthcoming years		
Large	Conceivable for specific markets only (e.g., luxury cars) due to high price and limited range		

Additionally, electricity onboard an electric car can be generated by a fuel cell. This technology has been investigated for decades, and production of small series of fuel cell vehicles (FCV) already started or is promised by carmakers to be released within the very next years. An FCV is an electric vehicle with a different energy storage compared to battery electric vehicles. It is equipped with a buffer battery, which is, however, much smaller compared to a BEV [7].

With respect to strategic and market perspectives as well as life cycle impacts of electric cars, their practicability in different segments of car sizes needs to be investigated (Table 1). Car size is most important in BEV since battery size must grow in parallel to the vehicle's weight.

In the next years, electric cars will be mostly small- or mid-size cars due to two main reasons: First, the weight limits the range of operation, which is a factor of suitability for daily use. Second, battery costs establish another main regulating factor: larger cars need bigger and much more expensive batteries.

On the contrary, PHEV and FCV are increasingly useful in the segment of medium-size and large cars because only a small fraction of energy is needed to be carried in the battery. The energy density of compressed hydrogen is close to fossil fuels, very much in contrast to the energy density of available batteries.

9.2.2 TECHNICAL COMPONENTS OF AN ELECTRIC CAR

According to Larminie and Lowry [9], the main components of a BEV can be divided into the electric battery, the electric motor, and a motor controller (Figure 1). The technical structure of a BEV is simpler compared to ICEV since no starting, exhaust or lubrication system, mostly no gearbox, and sometimes, not even a cooling system are needed.

The battery charges with electricity either when plugged in the electricity grid via a charging device or during braking through recuperation. The charger is a crucial component since its efficiency can vary today between 60% and 97%, wasting 3% to 40% of the grid energy as heat. The motor controller supplies the electric motor with variable power depending on the load situation. The electric motor converts the electric energy into mechanical energy and, when used within a drivetrain, to torque. In series BEV produced so far, central engines have been used; however, hub wheel electric engines are also possible and would be available for mass production (summarized in [5]).

Modern, highly efficient electric motors are based on permanent magnetic materials from which the strongest are alloys containing the rare earth elements (REE) neodymium and samarium, respectively. Usual alloys are both NdFeB and SmCo magnets [10].

This has caused some concern since REEs are scarce, and their export is controlled by a few countries, mainly by China (Helmers, unpublished work). However, electric motors for BEV do not necessarily contain REE. There are several types of electric motors, usually divided into alternating current (AC) and direct current (DC) types. There are both AC and DC electric engines built with and without permanent magnets, according to individual use.

In electric cars, traction motors without magnets are quite usual since they are cheaper (Loehr C, personal communication). A subspecies of AC motors are induction motors using no REE. The Tesla Roadster is equipped with an induction motor without REE, as will be the forthcoming Tesla Model S and the Toyota RAV4EV. In a more detailed view, it can be stated that there are several electric engines available operating without REE magnets: conventional mechanically commutated DC machines, the asynchronous machines, the load-controlled synchronous machines with electrical excitation, and the switched reluctance motors (Gerke W, personal communication). This gives the motor industry some flexibility.

9.2.3 BATTERIES FOR ELECTRIC CARS

It is still possible and useful to equip electric vehicles with lead-acid batteries. Cars of the Californian interim electric vehicle boom in the 1990s were partly driven by lead batteries (Table 2), nevertheless already offering a driving performance comparable to ICE cars. Today, for example, there are small electric trucks commercially available and equipped with lead batteries and a capacity of 13 to 26 kWh, allowing a maximum range of up to 200 km and a maximum speed of 60 km/h (numbers taken from a prospectus of Alkè Company, Italy, 2010). Also, a certain share of today's electric cars (e.g., by the Indian company REVA) are equipped with Pb batteries. In order to diversify the future battery technology and materials, it would be useful to keep Pb traction batteries for certain applications. Electric cars for smaller ranges, as e.g., in-town driving, so-called neighborhood electric vehicles, will be much cheaper if they are operated with lead-acid batteries instead of a lithium-ion battery. Additionally, there are recent performance improvements of the lead battery, thanks to a gel matrix and gassing charge [11].

However, the enormous increase in energy density offered by Li-ion batteries is the prerequisite for the expected widespread electrification of cars. Nickel metal hydride batteries were used in the interim time when the re-electrification of the automobile started in the 1990s. However, they do not offer enough power and have a worse environmental impact compared to Li-ion batteries (see below). The only alternative to Li-ion batteries with comparable power, the Zebra cell (Table 2), is based on molten salt and, thus, only useful for continuous every day use. Today, a lot of different Li chemistries are available, and prices are continuously decreasing for Li-ion batteries (e.g., summarized in [12]). However, the price for a complete Li-ion cell set offering 14 kWh capacity, allowing a 100-km electrical range of a small-size car (like a Smart, see below), is still in the order of 5,000 Euro including taxes. Life cycle impacts of the various Li-ion chemistries differ significantly (see below).

9.3 ENVIRONMENTAL IMPACT OF ELECTRIC CARS

9.3.1 WELL-TO-WHEEL EFFICIENCY OF ELECTRIC CARS

Considering the enormous worldwide increase of mobility expected for the future, the reduction of automobile energy demand is one of the most important challenges. In order to evaluate the technologies available, energy consumption is divided into the well-to-tank (WTT) and the tank-to-wheel (TTW) demands. WTT refers to the stage from the extraction of feedstock until the delivery of fuel to the vehicle tank [15]. TTW quantifies the performance of the drivetrain. Together, both result in the overall well-to-wheel (WTW) efficiency. The WTW evaluation allows estimation of the overall energy and efficiency of automobiles powered by different propulsion technologies (Table 3).

TABLE 2: Important materials commercially used in traction batteries for electric vehicles since 1998

Battery type	Active chemical components	Energy density (Wh/kg)	Costs (Euro/kWh)	Cars (examples)
Lead-acid	Pb/PbO_2	30 to 35a	up to 100d	GM EV1 (1996 to 1999), REVA
	H_2SO_4	10 to 40b	100 to 150a	
Nickel metal hydrides (NiMH)	various alloys, as e.g., LaNdNiCoSi	60 to 70a	300 to 350a	Toyota RAV4EV-I (1997-2003), Toyota Prius I-III
80c				
50 to 105b				
Zebra	NaCl-Ni	150c	500e	Th!nk City, Smart EV, Smith Electric Vehicles
80 to 120b				
Lithium-ion	Li-Ni-Co-Al	all types of chemistry: 120 to 150a	500 to 750f	Th!ink City, Mitsubishi I-MiEV, GM Volt/Opel Ampera, Nissan Leaf, Tesla
	Li-Ni-Mn-Co	500 to 1,000g		
	$LiMn_2O_4$			
	Li-titanate	85 to 160b		
	$LiFePO_4$	370h (cells only)		
	Li-Polymer (LiPo)			

Considering the fact that cars (light duty vehicles) are so important for worldwide public and private transport, it is astonishing that there are only two technical alternatives to the established ICEV available in the market: battery electric cars and hydrogen-powered fuel cell cars (Table 3). Cars equipped with hydrogen-powered fuel cells, however, are not yet available as series products, but manufacturers like Mercedes-Benz and Toyota promised to be close to releasing or have already released a small series of FCV. The main advantage of a FCV compared to a BEV is a much bigger range and quick refilling of the tank. However, the necessary H2 filling station infrastructure is available nowhere in the world, not regarding some single filling stations in a few city centers allowing regional mobility of hydrogen-powered fuel cell vehicles, which must return daily to the same filling station.

TABLE 3: Energy efficiency of the propulsion technologies available to the market (in percentages)

Propulsion technology	WTT (%)	TTW (%)	WTW (%)	WTW calculated (%)
ICE				
Petrol (gasoline)	79a, 86b	16a, 23k	10o, 13a, 12 to 14p, 14q, 20r	13 to 20
Petrol (gasoline) full electric hybrid	79a, 86b	30k, 37a	15o, 17 to 22p, 29a,q	24 to 32
Diesel	76c, 82a, 84b	23a, 28k	13o, 16 to 18p, 19a, 25r	18 to 24
LPG (propane + butane)	88d, 93e	16e	11o, 15e	14 to 20
petrol ref. + 6l				
CNG (methane)	65 to 86f, 85e	16e	12o, 14e, 21r	10 to 22
petrol ref. + 9m				
FCV				
H$_2$ fuel cell (gaseous H$_2$ stored in pressure tanks)	37g, 40c, 53h	50a, 56k	20 to 23p, 22a, 23s, 29q	19 to 30
BEV				
Electric car (literature)		73k, 80 to 90n	59 to 80t, 74k	
Electric car minimumi	15i	73k		11
Electric car optimumi	59 to 85j	90n		

9.3.2 EFFICIENCY UNITS

Efficiencies of different propulsion technologies may be expressed either by CO_2 equivalent emissions per course unit (e.g., CO_2/km), by energy units (MJ/km), or by percentages looking at the energy transformed to motion. Since petrol (in US: gas), diesel, LPG (propane + butane) as well as natural gas (methane) are all hydrocarbons and burn to mainly CO_2 while releasing energy, the consumed energy and the CO_2 emissions are proportional. If WTW data are expressed in energy units or CO_2 emissions, they may allow assessing different technology alternatives at least within the ICE sector. Often, these data include both the fuel chain and the operation of cars (e.g., [22]). However, comparing WTW data of ICEV with alternative technologies is usually complicated by the lack of data and testing schemes for alternative technologies. Good (realistic) data of one technology compared with bad (unrealistic) data regarding the alternative technology can fundamentally change the results of the efficiency evaluation or, following that, the life cycle assessment (LCA) comparison. We decided to review efficiency percentages (Table 3) of the available propulsion technologies for greater transparency. This way, the wrong impression of higher accuracy than available from the data, as well as erroneous conclusions, is avoided while comparing data of ICEV with alternative technologies.

9.3.3 INHOMOGENEOUS DATABASE

The available database is enormous but should be regarded inhomogeneous and, for some parts, as questionable. Some studies are based on modeled data (virtual cars), some on laboratory measurements with isolated engines, and some based on unified test schemes about actually driving the car. Some measurements were performed on laboratory test stands, others on real streets. They are also based on cars with various curb weights. Accordingly, there are strongly deviating results: Sullivan et al. [30], for example, resumed that, by 2015, diesel-driven cars will (still) have 14% to 27% less CO_2 emission due to a higher TTW efficiency, compared to petrol-fuelled cars and based on German certification data. But in reality, consumers preferred bigger motorization than modeled. A comparison by a German non-governmental organization revealed that certified CO_2 emissions of new diesel cars on German streets increased since 2000 and coincided with falling CO_2 emissions of new petrol-fuelled cars at 173 g CO_2/km in 2006 (reviewed by Helmers [1]).

9.3.4 STANDARDIZED DRIVING CYCLES: CONTRAST TO REALITY

A major problem in well-to-wheel efficiency studies is that most data are based on artificial test procedures, which are also different from one region of the world to another. The German Ministry of Transport, Building and Urban Development recently demonstrated [31] that the majority of cars consume around 25% more fuel and thus emit more CO_2 than certified. Most of more than 100 cars investigated are within 40% of excess, while a few percent of the vehicles in this spot check revealed fuel consumption up to 70% higher than certified due to the European test scheme [31]. A worldwide unified test scheme is therefore currently under international negotiation [31]. If this unified driving cycle includes alternative propulsion technologies, it could serve as a basis to generate more reliable TTW data in the future.

9.3.5 INTERNAL COMBUSTION ENGINE VEHICLES

ICEVs are powered by petrol (gasoline), diesel, propane and butane, or methane (Table 3). These fossil hydrocarbons may be substituted or exchanged with biofuels like bioethanol, biodiesel, or biomethane, which is not the subject of this paper. An integrated greenhouse gas (GHG) assessment of the process chains of natural gas and industrialized biomethane is provided by Arnold et al. [32], revealing that GHG emission data will change in the future due to the development of new sources and markets like, e.g., the growing market for liquefied natural gas. WTT data for petrol (gasoline), diesel, propane and butane, and methane provisions reveal that up to 24% of the contained energy is already being consumed within the chain. In the exceptional case of methane, pumped trough up to 7,000 km from Siberia to Europe, there is a WTT loss of up to 35% (Table 3).

In general, tank-to-wheel efficiencies of ICEV are very low with 10% to 25% because 75% to 90% of the energy is lost as heat instead of propelling the car. However, ICEV has been successful on the market for more than a century due to the very high energy density (up to 20 times higher compared to Li-ion batteries) of carbon-based fuels available world-

wide since over 100 years for low prices. The electric hybrid ICE concept brought an efficiency jump; TTW's of hybrid ICE roughly doubled
compared to that of ICEV without electric assistance (Table 3). Toyota, in
1997, introduced this technology to the market in a large-scale production,
followed by Honda in 1999 (reviewed in [5]).

ICEV powered by gaseous hydrocarbons, namely methane, and propane and butane, are similarly inefficient like petrol-powered cars. However, if the higher caloric value is used for achieving higher compression,
gas-driven propulsions can be as efficient as the diesel engine (+6% to 9%
TTW, see Table 3). Unfortunately, car manufacturers do not utilize the
potentials of gaseous propulsions so far. Altogether, WTW calculations
reveal that, during the operation of ICEV, between 68% and 90% of the
entire energy is wasted (Table 3).

9.3.6 FCV AS ALTERNATIVE PROPULSIONS

Within the two technical alternatives available, H2-powered fuel cell vehicles reveal a problem, which is losing 50% to 60% energy each during
fuel production (here both from hydrogen production by steam reforming
as well as by water electrolysis) and during fuel cell operation and driving.
Altogether, WTW efficiency of FCV seems to be nearly as low as the efficiency of ICEV, with 19% to 30% (=70% to 81% energy loss). However,
in terms of CO_2 emissions, several authors see WTW advantages of FCV
over ICEV (e.g., [22]); see below.

9.3.7 BEV ANALYSIS IN TERMS OF WTT, TTW, AND WTW EFFICIENCY GENERATES A COMPLEX PICTURE

First, electricity production can take place under very different conditions:
When electricity is generated from fossil sources in an inefficient power
plant and loaded with an inefficient charger to the BEV, up to 85% of the
energy may be lost, resulting in a WTW efficiency of only 11% comparable to petrol-operated ICEV (Table 3). Actually, in a pre-series BEV, car
chargers with only 60% efficiency have been implemented. Today's most

efficient battery chargers, however, have efficiencies of up to 97%. Today, chargers are mostly implemented in the car, while in the California BEV of the 1990s, they were partly not integrated in the cars, complicating the comparability of older WTW numbers.

Today, WTWs reported for BEV are between 59% and 80%. This high efficiency is due to the fact that only very little energy is wasted in the drivetrain (Table 3). Altogether, BEV represent the only alternative technology offering an efficiency jump in individual mobility, consuming up to four times less energy than today's cars. This has been confirmed during measurements on a car converted from combustion engine to electric (see below).

However, BEV should be primarily and increasingly loaded with electricity from renewable sources and must be equipped with an efficient charging unit (Table 3). Although being much more efficient so far, BEV cannot cover all mobility needs due to range restrictions, so the other technologies (ICE and FCV) are still needed under sustainability-optimized conditions.

9.3.8 FROM EFFICIENCY TO ENERGY UNITS

Consumption of the competing technologies in terms of energy units should also be known. Ideally, data should be taken from the same car as an ICE as well as an electrical version, which is unfortunately not available on the market. Therefore, in 2011, a Smart car has been converted from petrol to electric in the laboratory of the authors, revealing realistic data (technical specifications see below). This car consumed 5.3 L petrol/100 km, as quantified during a test cycle under standard road conditions with an original 40-kW engine prior to the replacement of the propulsion, which can be converted to 1.67 MJ/km. The test cycle was based on a 46.8-km route consisting of villages (20%, 50 km/h), autobahn (24%, 100 km/h), and country roads (80 km/h). The same test cycle was driven before and after electric conversion, both at air temperatures of 15°C to 20°C, without heating (no A/C available); the car has been weighed each time. After installing a 25-kW electric engine, the electrified Smart had better acceleration and consumed 14.5 kWh/100 km on the above test cycle, including charge losses, which are equivalent to 0.5 MJ/km. This number is reflected in literature: Majeau-Bettez et al. [33] based his LCA calculations on a BEV electricity consumption of 0.5 MJ/km. Since the weight of the Smart increased by 22.2% due to the added Li battery (additional information given below), the weight-normalized energy consumption of the electrified Smart is 0.4 MJ/km. This realistic experiment fairly reveals the fourfold energy efficiency advantage of electric cars proposed in literature and reviewed in Table 3: ICEV operated with petrol has only 16.5% (13% to 20%) TTW energy efficiency, while electric cars today are believed to achieve 65% (53% to 77%) WTW, which is a four-times higher energy efficiency (Table 3).

It is of course more difficult to get street data for FCV. The first realistic hydrogen consumption data of pre-series cars are reported by car magazines, revealing, e.g., 1.11 kg of 700 bar H2/100 km for a Mercedes F-cell B-class car (pre-series model) with a curb weight of 1,809 kg [34]. The H_2-consumption of a Honda FCX Clarity is only slightly higher, considering its lower tank pressure [35].

Due to the German Hydrogen and Fuel-Cell Association [36], hydrogen of 700 bars pressure contains 1.3 kWh/L of energy (specific gravity 40.2 kg/m3). This way, energy consumption of the B-class fuel cell car can be converted to 35.9 kWh/100 km or 1.3 MJ/km, respectively. Converted to the curb weight of the electrified Smart (880 kg), only 0.65 MJ/km would be spent by a FCV of the same weight, which is about one third more than the electric car consumes (0.5 MJ/km). This also matches the energy consumption of a FCV vehicle modeled by Simons and Bauer [37], revealing 0.68 MJ/km (converted to a vehicle's weight of 880 kg). Linssen et al. [38] simulated between 0.95 and 1.57 MJ/km for small and large FCV vehicles (860 to 1,270 kg weight), respectively, which are based on compressed hydrogen technology.

A B-class car of today with ICE engine extrapolated to the same weight (1.8 t) would consume 11.6 L/100 km (certified fuel consumption taken from manufacturers specification + 25% reality supplement, and extrapolated to 1.8 t) equivalent to 3.7 MJ/km, 2.8 times the fuel cell version. The TTW-proportion FCV/ICE (petrol) is 2.7, according to the literature (Table 3). In conclusion, energy efficiency of a fuel cell car seems to be not far away from the BEV. However, the WTT efficiency of pressurized hydrogen is bad since up to 63% energy is lost within the delivery chain (Table 3) This led to criticism about driving with hydrogen as fuel (e.g.,

[28]). However, under unfavorable conditions, efficiency of electricity provision to a BEV can be even worse (Table 3).

9.4 GREENHOUSE GAS EMISSIONS OF ELECTRIC CARS IN OPERATION

9.4.1 BATTERY ELECTRIC VEHICLES (BEV)

The greenhouse gas emissions of the BEV can easily be calculated based on its electricity consumption and following the GHG emission associated with the local electricity production [1,5]. In the literature, there are plenty of data quantifying greenhouse gas emissions of electric cars in operation (e.g., [39]).

The crucial number in this context is the electricity consumption of electric cars under street conditions. In a review evaluating 21 studies from 1999 to 2009 (18 studies therein from 2007 to 2009), BEV and PHEV consumed a mean of 17.5 kWh/100 km [29]. In contrast, in a few studies focused on LCA quantification, a much higher electricity consumption has been supposed: For example, Helms et al. [40] based their modeling on an electricity consumption of 20.4 kWh/100 km in urban areas, 20.8 kWh/100 km in extra-urban areas, and 24.9 kWh/100 km on the highway, respectively. Pehnt et al. [41] summarized 21 to 24 kWh/100 km. Held and Baumann [42] based LCA quantifications on electricity consumptions of 18.7 kWh/100 km for a mini-class BEV (736 kg curb weight, size of Smart, see below) and 22.9 kWh/100 km for a compact-class BEV (specified curb weight 1,115 kg). In some cases, the assumed electricity consumption of a standard BEV seems to be quite high since only mid-size cars equipped with heavy batteries are taken for calculation (e.g., by Helms et al. [40] based on a BEV with 1,600 kg curb weight).

On the contrary, we assume that, in the next few years, electric cars will be mostly small or mid-size cars (see above). An e-conversion project, performed in the laboratory of the authors in 2011, may illustrate weight data exemplarily: A Smart built in the year 2000 has been e-converted from combustion engine to an electric car. A 25-kW electric engine and a 14-kWh LiFePO$_4$ battery were installed, allowing a range of up to 120 km. The curb weight grew by 161 kg to 880 kg, which is little more than half of the car weight assumed by Helms et al. [40]. A Mitsubishi I-MiEV weighs 1,110 kg and is equipped with a 16-kWh Li-ion battery [43]. The Nissan Leaf, however, the first high-volume mid-size electric car, weighs 1,525 kg and operates a 24-kWh lithium battery [44].

Furthermore, BEV electricity consumption data underlying LCA modeling and indicating, e.g., a yearly CO_2 emission of a car fleet are necessarily more or less theoretical today (e.g., [8]). Here again, consumption data derived from certified driving cycles are simulating much lower fossil fuel consumptions for ICEV than realistic and, on the contrary, probably higher than realistic for BEV, as established in the following. We suggest that street condition fuel consumption should be compared for both BEV and ICEV, including charging losses for BEV of course. Only a few real-life electricity consumption data are available so far, which usually cannot be found in scientific literature. The Smart Fortwo converted to an electric car in the laboratory of the authors consumes 14.5 kWh/100 km on the above described test cycle. A Mitsubishi i-MiEV, the first high-volume electric car on the market, consumes 16.94 kWh/100 km on the street [45]. As expected from an electric car, consumption is lowest inside the city, where ICEV conversely exhibit highest fuel consumptions. The Mitsubishi I-MiEV demonstrated the following electricity consumption data during ADAC-testing: 11.3 kWh/100 km (urban), 15.0 kWh/100 km (extra-urban), and 24.6 kWh/100 km on the autobahn at higher speed, respectively [45]. This highlights again the need for a critical data evaluation: BEV will replace ICEV mainly in the local urban area or within local regional traffic, generating possibly higher CO_2 advantages than seen in the data derived from standard cycle consumption of BEV and ICEV. Interestingly, Mitsubishi publishes 13.5 kWh/100 km due to the ECE R101 cycle [43], which confirms the independent test results (cited above) for urban and extra-urban consumption.

On the other hand, electricity consumption specified by manufacturers of BEV can be too low to be realistic. For example, Tesla motors claims its 1,230-kg roadster consumes only 11 kWh/100 km [46]. Nissan Leaf is the second high-volume BEV available on the market and the first series BEV specially developed for electrical driving. USEPA certifies electricity consumption of the BEV Nissan Leaf (curb weight 1,525 kg [44]) as

19.9 kWh/100 km in the city and 23.0 kWh/100 km on the highway, respectively [47]. Unlike the European driving cycles, EPA certifications are claimed to be more realistic since they contain 'faster speeds and acceleration, air conditioner use, and colder outside temperatures than usual until 2008' [47]. Nissan itself specifies electricity consumption of its Leaf as 17.3 kWh/100 km [44]. A Mercedes A-class pre-series electric car has been tested, resulting in 19 kWh/100 km [48]. In conclusion, an electricity consumption of up to 20 kWh/100 km should be realistic for a European mid-size car moved in urban areas or extra-urban at limited speeds, which is the favorite use of electric cars. The 20 kWh/100 km electricity consumption has also been confirmed by a scientific review process (personal communication, [HJ Althaus, 2011]).

Accordingly, we can assume a realistic electricity consumption of 15 to 20 kWh/100 km for urban and extra-urban traffic. European small-size cars (like Smart or I-MiEV) will be located in the lower end, while European mid-size cars (like Nissan Leaf) will be found in the higher end of this range. Large-size BEV and BEV driven at higher speeds (autobahn) can be expected to consume more than 20 kWh/100 km. However, in the forthcoming years, large-size BEV will probably not be the dominant application within the sector of electric cars.

The lower the carbon impact accompanying the electricity production in a country, the lower is the greenhouse gas emissions of the BEV in operation. However, countries such as Australia, China, India, Poland, and South Africa produce between 68% and 94% of their electricity by combustion of coal [50]. Coal represents 78% of China's electricity generation [51], resulting in 743 g CO_2/kWh (IEA, number for 2009). According to Yan and Crookes [51], Chinese coal-based electricity production generates CO_2 emissions in the range of 194 to 215 g/km operating a BEV, which is much higher compared to the 84 to 113 g/km of BEV operated in Germany under grid conditions (563 g CO_2/kWh [52, number for 2010]) and for BEV consuming 15 to 20 kWh/100 km.

However, operating a BEV in China anyhow can lead to significant GHG savings if compared to ICEV operated with Chinese coal-to-liquid (CtL) production fuel. For CtL of this kind, the carbon emissions are 717 to 787 g CO_2/km [51].

9.4.2 PLUG-IN HYBRID ELECTRIC VEHICLES

PHEV has an electricity consumption quite similar to BEV as long as they drive electric. Weight and cost savings due to a smaller battery are compensated by an additional combustion engine (also called range extender). Hacker et al. [29] reviewed eight studies on PHEV revealing a mean electricity consumption of 17.4 kWh/100 km. However, depending on the size of the battery and concept of the individual car, the electric range of PHEV is smaller than the range of BEV: The first high-volume PHEV, the GM Volt/Opel Ampera, has a 40- to 80-km range according to its carmaker [53], while the Toyota Prius plug-in (2012) will have an electrical range of around 20 km only. Electricity consumption in electrical mode and fuel consumption during operation of the combustion engine are known only from publications of the motor press and represent preliminary results allowing a rough estimation of their GHG emissions in operation: Auto Motor Sport [54] reported 23.5 kWh/100 km in the electrical mode as well as 6.7 L petrol/100 km in the ICE modus for the Opel Ampera. AutoBild [55] reported 22.6 kWh/100 km in the electrical modus versus 7.7 L petrol/100 km. A GHG impact compilation for a 100-km driving cycle of the Opel Ampera (60 km electrical range plus 40 km of driving with combustion engine, no charging underway) results in 159 g CO2/km based on the following assumptions: 563 g CO_2/kWh for German electricity (number for 2010 [52]), 2,310 g CO_2 due to the combustion of 1 L petrol [5], and supply chain emissions of 506 g CO_2/L petrol (calculated from Öko-Institut [56]). Carmaker Opel, however, specifies a value of 27 g CO_2/km [53].

A Toyota Prius plug-in consumes 3.4 L petrol/100 km in ICE mode according to an ADAC test [57] and 21.8 kWh/100 km in the electrical mode [58], respectively. The journal AutoBild [58] measured 3.8 L petrol/100 km in ICE mode. Regarding an electrical range of 20 km, this results in an emission of 106 g CO_2/km (calculated for 100 km). Carmaker Toyota specifies 59 g CO_2/km due to ADAC [57].

Over a 100,000-km lifetime (this number chosen in order to ensure comparability with LCA calculations shown below), an Opel Ampera would sum up 15.9 t CO_2, while a Toyota Prius plug-in comes to 10.6 t of CO_2, respectively, both charged under German mean electricity grid conditions. However, these lifetime balances for PHEV operation are based on the assumption that the battery is completely discharged every time and every trip continues with petrol. In reality, operational lifetime carbon footprints vary strongly with respect to individual use of the cars.

TABLE 4: CO_2-equivalent emissions due to the Li-ion battery production

	CO_2-equivalents kg/kWh battery	Battery chemistry investigated	Remarks
Notter et al. 2010, [61] Althaus 2011	52	$LiMn_2O_4$	Bottom-up LCA approach (Althaus 2011)
Ishihara et al. 2002 [63]	75	Li-Ni-Co and Li-Mn	Bottom-up LCA approach (Althaus 2011)
Zackrisson et al. 2010 [64]	166	$LiFePO_4$	Top-down modeling from producers (Althaus 2011)
Frischknecht 2011 [65]	134	Not specified	Top-down modeling from producers (Althaus 2011)
Majeau-Bettez et al. 2011 [33]	250	$LiFePO_4$	Industry data considered (remark by Majeau-Bettez)
	200	$LiFe_{0.4}CO_{0.2}-Mn_{0.4}O_{0.2}$	
Data range presented by five different groups on the 43rd discussion forum on LCA, Zurich, 2011 (reviewed by Frischknecht and Flury 2011) [67]	(66 to 291)a	diverse	-

These preliminary findings on the first two reality PHEV exhibit that efficiency and ecoimpact of PHEV will vary very much depending on the technical conception, which is expected to diversify. Moreover, it depends on the average distances traveled in daily use where the two propulsion modes are mixed together.

TABLE 5: Characteristics of electric cars (compact class) today in comparison with the converted Smart

	Held and Baumann 2011	Held	Lambrecht	Frisch-knecht	Althaus	Freire	This study	
							New Smart	Converted Smart
Car weight (kg)	1,037	1,670	N.a.	1,632	1,880	1,531	880	880
Lifetime car (km)	171,600	171,600	150,000	150,000	150,000	200,000	100,000	100,000
Battery weight (kg)	200a	400	250	312	400	329	160	160
Lifetime battery (km)	114,400	114,400	100,000	75,000	150,000	100,000	100,000	100,000
Electricity consumption (kWh/100 km)	18.7	22.9	22	20	20	18.8	14.5	14.5
Electricity mix	DE	DE	DE	CH	CH	PT	DE	DE
Climate change impact (g CO_2-eq/km)	(170)a	240	225	150	95	165	140	108

9.4.3 FUEL CELL CARS

Linssen et al. [38] have quantified the CO_2-equivalents of different supply paths of hydrogen based on natural gas as the hydrogen resource and including pressurization and transport. CO_2 emission of German H2 production would result in 96 g/MJ, while a Norwegian production would result in 83 g/MJ, respectively [38]. FCV energy consumption between 0.65 for small FCV (car weights see above) and 1.57 MJ/km for large FCV, respectively [38], results in 54 to 151 g CO_2/km. Eberhard and Tarpenning [46] published a consumption of 152 g CO_2/km for the Honda FCX operated in the USA. This order of magnitude was also confirmed by Höhlein and Grube [59], who also concluded that, for H_2 generated by electrolysis and powered by wind electricity, CO_2 supply path costs can be lower than 25 g/km. Simons and Bauer [37] calculated CO_2-equivalent costs of around 150 g CO_2/km (steam methane reforming from natural gas), 105 g CO_2/km (biomass in a steam methane reforming process of gasified wood), 320 g CO_2/km (H_2 electrolysis powered with the European grid electricity), or 80 g CO_2/km (H_2 electrolysis powered with the Swiss grid electricity), respectively. The modeling of Simons and Bauer [37] is based on a car weight of 1,434 kg. Wu et al. [60] accounted between 30 and 230 g CO_2/km (H_2 produced with wind energy vs. that from North American natural gas) for the operation of fuel cell hybrid vehicles. For a Chinese FCV, 146 g CO_2/km is supposed [51]. While the operational ecoefficiency of BEV is very much dependent on the electricity source, the CO_2 impact of FCV is strongly dependent on energy and hydrogen resources used for H2 production.

Assuming a supply chain resulting in 100 g CO_2/km emissions over 100,000 km, some 10 t of CO_2 is produced. This would be very similar to a Toyota Prius plug-in PHEV and only 22% more than the operational life cycle emissions of the electric Smart (see below), which is smaller and lighter than a FCV car.

9.5 LIFE CYCLE ASSESSMENT OF ELECTRIC CARS

In order to quantify the LCA of electromobility, the impacts of electric vehicle production, maintenance and disposal on the one hand, and the impacts of operation including fuel provision on the other hand are quantified [61]. Impact of road construction, maintenance, and disposal are neglected here since there are no differences between ICEV and BEV. LCA is usually calculated separately for the glider (or platform = vehicles without engine, transmission, fuel system, or internal combustion components of any kind), the drivetrain (electric engine and associated compounds, transmission, and charging infrastructure), the battery production, and the maintenance and end-of-life treatment, respectively [61]. Other studies also distinguish sub-parts like inverters/electronics, the generator, and other components [42].

The overall environmental impact according to the international standard ISO EN 14040 and 14044 [62] includes quantification of, e.g., the abiotic depletion potential, the non-renewable cumulative energy demand [61], the acidification potential [42] and, of course, the global warming potential as CO_2-equivalents. However, the global warming potentials of BEV production and use are discussed controversially in science and public, while the other criterions are found in the scientific debate only. We like to point out some of the critical details within the discussion and add preliminary data from a used car converted from ICE to electric.

9.5.1 ENVIRONMENTAL BURDEN OF THE LI-ION BATTERY PRODUCTION

There are only a few complete LCA studies available presenting detailed inventories [33]. Only two of them [33,61] are published together with comprehensive input data and model description fully available on the internet upon title research. Some further studies are published only with limited input data and are also reviewed in the following (Table 4).

Notter et al. [61] conclude that Li-ion battery plays only a minor role (between 5% and 15%) regarding the overall environmental burden of E-mobility, independent from the impact assessment method used. In contrast, Held and Baumann [42] calculated a global warming potential (GWP) of the battery production in the order of 5 to 10 t CO_2-eq, dominating the GWP LCA of mini- and compact-class BEV propelled with renewable electricity and, furthermore, accompanied by high and dominating acidification potential of 40 to 80 kg SO_2-eq for the whole battery.

However, Held and Baumann [42] supposed that mini- and compact-class BEV should have batteries of 20- to 40-kWh capacities. In contrast, the authors' converted Smart has an electrical range of more than 100 km with a 14-kWh Li-ion battery, sufficient for regional mobility. Compact-class BEV Nissan Leaf is said to have a cruising radius of 175 km with a 24-kWh battery [66]. In conclusion, battery sizes required for BEV operation seem to be partly overestimated in the literature so far (see also Table 5).

Not surprisingly, environmental costs of battery production and usage are subject to intensive scientific discussion, revealing corresponding CO_2-equivalent emissions of a great variety (Table 4). However, these variations are only to a little extent due to the battery chemistry and, least of all, due to the metal lithium since a Li-ion battery contains only about 1% lithium or 80 g Li per kWh energy content [68]. Also, the Li purifying process is not energy-intensive [33], nor is Li related to a comparably high depletion of resources, according to Althaus et al. [68]. The $LiFePO_4$ battery used in the Smart conversion project performed by the authors contains 3.4% Li. Besides the Li, the $LiFePO_4$ battery (manufacturer: Calb, China) has a content of 42% Fe, 16% P, 5% graphite, 3% C, 6% Al, and 10% Cu, respectively (MSDS accreditation certificate, 2009). The components of the highest relevance within the whole battery LCA are the anode and cathode materials graphite, copper, and aluminum [61,68]. Majeau-Bettez et al. [33], however, identified battery and components manufacturing, as well as the positive electrode paste, as being the most GWP-intensive components.

Notter et al. [61] primarily evaluated the environmental burden of a $LiMn_2O_4$ battery and also found that two other of the often used active materials exhibit only a small increase of 12.8% ($Li-Mn-Ni-Co-O_2$) and a decrease of 1.9% ($LiFeO_4$), respectively, in environmental burden (EI99H/A). This is confirmed by Majeau-Bettez et al. [33], quantifying only a 25% difference in the GWP during the production of the two different materials investigated (Table 4). Instead of the Li battery chemistry, methodical differences in LCA quantification seem to cause data deviations [49]: While the 'bottom-up' method of, e.g., Notter, Althaus, and Ishihara leads to lower CO_2-equivalent emissions, a 'top-down' approach (Zackrisson and Frischknecht) seems to generate higher results (Table 4). In addition to the different approaches to quantify costs of Li battery production, the application scenario has to be defined, which is a theoretical attempt since there is no widespread use of electric cars yet in everyday life: What is the annual distance traveled with an electric car? How many years will an electric car be in operation? How long does a battery in a car last? The latter depends on battery chemistry, its cycle strength, the quality of production, and pattern of everyday use. Also important is how charging is performed in everyday life - quick (at higher currents) or slow - and to which temperature fluctuations the battery will be exposed. Statistically usable data are not expected before BEV has been brought on the streets for routine use.

In conclusion, LCA database, so far, for Li-ion battery production still seems to be in substantial movement. This is even more the case when electricity sources change from fossil to renewable. Tao et al. [69] claim that CO_2 emissions from electricity consumption during Li-ion battery production can be reduced by 95% to 98% if the production site is shifted from China/Europe to Iceland with its geothermal energy resources. Electricity production in Iceland causes a footprint of 18 to 23.5 g CO_2/kWh only [69].

9.5.2 CO_2-LIFE CYCLE IMPACT OF THE CONVERTED SMART (BEV VS. ICEV)

Carbon footprint and environmental impact quantification of standard automobile parts can be expected to generate less volatile data than the Li-ion battery. The glider, a car minus motor, gearbox, and fuel equipment, is taken as a useful basis for modeling [61]. A full LCA of automobile use includes an impact quantification of the glider's production, the manufacturing of propulsion components (in an ICEV—the combustion engine, gearbox, and fuelling system; in a BEV—the electric engine plus electric controller system) rather than, for a BEV, the footprint of battery production. CO_2 emission during operation of the car is quantified considering the consumed carbon-based fuel (ICEV) or indirect CO_2 emission during electricity production (BEV). Even if a detailed LCA modeling of an individual car is not available for all its technical parts, a simplified life cycle assessment can be performed with comparative calculations based on LCA models published in detail: As an example, lifetime carbon footprints were estimated here according to the conversion of a used Smart in the laboratory of the authors (Figure 2). The LCA for the glider, drivetrain, and battery of the Smart were recalculated based on the data by Notter et al. [61]. Measured weights of the Smart's glider, the gearbox (kept), the new electric motor, the battery as well as the controller, and further accessory parts were converted relative to the carbon footprint data by Notter et al. [61]. The detailed material composition of the glider was not considered in this assessment. However, unlike the LCA data published so far, this estimation is not based on modeled cars but on a used car purchased on the market and then converted to electric. Also, different to most of the published data, petrol and electricity consumptions of the same car were measured prior to and after the technical conversion: The Smart has been driven along the same route and the same street conditions before and after electric conversion.

Real street data were chosen to evaluate the energy consumption. The ICEV Smart belongs to cars with some of the highest deviations between certified and reality fuel consumption (up to 60% [1]). We chose a mileage of 100,000 km on the assumption that the first battery will be kept within this range. Also, fossil-fuelled Smarts, on the other hand, can be expected to drive between 60,000 and 120,000 km with the first engine according to our knowledge, which allows a comparison between the ICEV and the electric version over 100,000 km. Operation emissions are both dominating the CO_2 life cycle of the fossil fuel-propelled Smart (column 1, Figure 2) as well as an electric Smart (column 2, Figure 2) if charged with German mean grid electricity. This is in accordance with the findings of Helms et al. [40] as well as Notter et al. [61]. Other studies report higher BEV CO_2 life cycle impacts during operation when charged with grid mix, e.g. Helms et al. [40] with approximately 20 t (mid-size car) and Held and Baumann [42] also with approximately 20 t (mini-class), however recalculated for 150,000 to 171,600 km of operation. Other differences will partly be due to a higher battery and car weight (Table 5). However, we calculated the direct CO_2 emissions due to fuel combustion (including fuel chain emissions) as well as indirect CO_2 emissions due to consumption of the German grid electricity (for details, see explanatory notes in Figure 2). Calculation of CO_2-equivalents based on ecoinvent 2.2 database adds up to 9.4 t instead of 8.2 t for the operation phase of the electric Smart (column 2 in Figure 2).

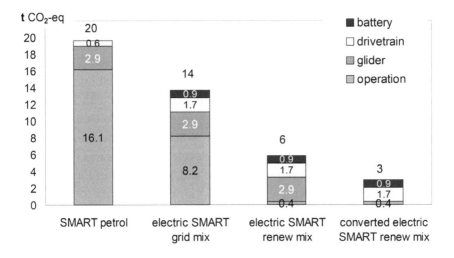

FIGURE 2: CO_2 life cycle assessment based on a converted Smart car. Preliminary and simplified CO_2 life cycle assessment based on data obtained during a conversion of a used Smart from an ICEV to a BEV. Calculation conditions: 100,000 km lifetime, grid electricity 563 g CO_2/kWh (German Federal Environmental Agency, number for 2010), and renewable electricity mix 30 g CO_2/kWh (reviewed in [5]). Smart I model, petrol-fuelled, built in the year 2000 (also calculated as being 'new'), and purchased in 2011 with 106,000 km driven. ICE engine, tank components, and exhaust system were removed, gearbox was kept. Fuel chain emissions of 2.88 t (included in 16.1 t operation emissions) calculated according to Öko-Institut [56] for petrol provision. CO_2 impacts of the battery, drivetrain, and glider production calculated according to Notter et al. [61]. Weight of the glider, 597 kg. Cutoff allocation rule applied for the glider in column 4 (converted Smart). Electric engine including gearbox and all accessories weigh 122 kg; battery weight is 161 kg. Further information is provided in the text.

9.5.3 LITERATURE COMPARISON OF THE SMART
E-CONVERSION CARBON FOOTPRINT

According to our simplified LCA model, life cycle CO_2 emissions could be reduced by 30% when switching from the petrol version to a new electric Smart (Figure 2). Fuelling the electric Smart with renewable electricity mix causes an additional 55% reduction of life cycle CO_2 emissions (second and third column in Figure 2). However, we have chosen an already driven Smart with combustion engine over 100,000 km and converted it, so we can omit expenditure for the glider production (cutoff rule); this way halved again the life cycle CO_2 emissions and ended up with 2.8 t (from column 3 to column 4 in Figure 2). Altogether, life cycle CO_2 emissions when driving a Smart have been reduced by 86% due to this model (Figure 2, from the first to the last column). Life cycle CO_2 emissions for the new electric Smart (first column in Figure 2) and, not shown in Figure 2, that of a converted used Smart propelled with German electricity mix (10.8 t in lifetime) are selected to calculate the overall g CO_2/km impacts in Table 5.

Other findings in Table 5 need to be commented: Electric Smart has the lowest climate change impact, which was expected since it is the smallest car. It confirms a finding already published [49]: The smaller the BEV, the more ecoefficient it is. This is also known from ICEV; however, it is more pronounced within BEV due to the high battery weight.

The data published by Althaus and Frischknecht (Table 5) seem to be in compliance with the Smart numbers. However, Althaus and Frischknecht modeled BEV twice the Smart's weight and even heavier. On the other hand, they based their life cycle emissions evaluation on Swiss electricity consumption. Swiss electricity GHG intensity is around 140 g CO_2-eq/kWh (number for 2005 [37]), while in Germany, 563 g CO_2/kWh (number for 2010, see above) must be taken into account. Lambrecht, Held as well as Held and Baumann [42] published larger climate change impacts. This is evident particularly for data extracted from Held and Baumann [42] based on a mini-class car quite comparable to the Smart. Since the carbon footprint we calculated for the Smart and Held and Baumann [42] calculated for a mini-class car (around 3 t CO_2) for the platform (glider) is nearly the same, the deviation is due to a higher electricity consumption of 18.7 kWh/100 km assumed by Held and Baumann [42]. Also, more than 6 t CO_2 was modeled by Held and Baumann [42] for production and maintenance of the battery, which is six times more than we assumed based on the data by Notter et al. [61].

Charging the Smart with electricity of renewable origin (30 g CO_2/kWh), however, would considerably decrease the overall climate change impact; the climate change impact of a new electric Smart would be more than halved on 63 g CO_2/km, and the impact of the converted Smart is down to 30 g CO_2/km, each calculated for 100,000 km of operation (not shown in Table 5).

9.5.4 LIFE CYCLE IMPACT OF PLUG-IN HYBRID ELECTRIC VEHICLES

Samaras and Meisterling [70] published ICEV life cycle CO_2-equivalent costs in the order of 270 g/km and for HEV of 180 to 190 g/km, respectively. Interestingly, with increasing the battery size of PHEV, CO_2-equivalents decrease only little in the order of a few grams per kilometer [70]. Similarly, Helms et al. [40] reported life cycle CO_2-equivalent emissions for PHEV in the order of 150 g/km, operated with German average electricity (150,000 km life cycle). Their data revealed advantages for BEV over PHEV only in combination with wind power [40]. Due to high battery production costs assumed, Held and Baumann [42] found advantages of PHEV over BEV in the criterions GWP and acidification potential (SO_2-equivalent), with the latter not being confirmed by Helms et al. [40]. Also, Althaus [49] stated that 'electric vehicles with sufficient battery capacity for normal use and a range extender for special use perform better than electric vehicles with larger batteries'.

9.5.5 LIFE CYCLE IMPACT CATEGORIES OTHER THAN GLOBAL WARMING POTENTIAL

Life cycle categories further than GWP implemented in LCA reports on electric cars published so far are abiotic depletion potential and non-renewable cumulative energy demand. Notter et al. [61] aggregated the three categories by the Ecoindicator 99 method, concluding that the environmental burden of BEV is significantly lower compared to ICEV. Aggregation to Ecoindicator 99 method revealed an environmental impact of BEV around 37% below that of ICEV. Althaus [49], however, pointed out additionally that BEV charged with electricity with a significant portion of nuclear energy will be associated with a backpack of nuclear waste production. Also, he identified disadvantages of BEV compared to ICEV in the land use category and, particularly noticeable, in the human toxicity potential category, both due to the production of the lithium battery. On the other hand, there are advantages of BEV over ICEV in the impact criterions resource damage and photochemical oxidation potential [49]. Also, Held and Baumann [42] found out about a distinct disadvantage of BEV in the acidification potential. This is due to the sulfur emissions during the smelting of metals like Cu and Ni and may also be due to associated S_2 emissions when components are produced in countries like China, where electricity production is dominated by coal-fired power plants. However, there are differences within the batteries: LiFePO4 batteries have a lower acidification potential since they contain no nickel [71].

Helms et al. [40] did not report elevated acidifying emissions accompanying the BEV production but emphasized the SO_2 emissions during BEV operation caused by electricity production from coal. This was confirmed by Helmers [1], who calculated the SO_2 emissions accompanying the power usage of a BEV in Germany based on year 2006 data, revealing up to tenfold higher indirect SO_2 emissions of a BEV compared to ICEV direct SO_2 emissions. Indirect BEV emissions of NOx and fine dust according to electricity production, however, were smaller than direct ICEV emissions [1]. Majeau-Bettez et al. [33] quantified the LCA emissions of battery production and use, reporting 14 to 19 g CO_2-eq/km (battery only) for two Li-ion batteries (specified in Table 4). They quantified a lot of additional impact categories (freshwater and terrestrial ecotoxicity, freshwater eutrophication, marine ecotoxicity, metal depletion, ozone depletion, and particulate matter formation) to those mentioned above and concluded that Li-ion batteries are environmentally favorable compared to the NiMH battery [33].

9.6 CONCLUSIONS

The electric car seems to be a suitable instrument and a sustaining measure towards a more sustainable mobility future since it is four times more energy efficient compared to ICEV. Therefore, it is seen as a milestone towards a 'Great Transformation' [4]. The TTW efficiency advantage of BEV over ICEV, together with the efficiency jump by Li-ion batteries, enable the electrification of the automobile as long as it is moved in regional ranges of up to 100 km per day. However, WTW efficiency of electric cars can reach exemplary figures only when electricity is provided by very efficient power plants and infrastructure, best with renewable energy production. Also, electric cars should be incorporated into a variety of modern mobility concepts (e.g., [72]).

Energy efficiency of an FCV propelled with hydrogen is only slightly lower compared to BEV; however, a lot of energy is lost during production and provision of compressed H_2 even in the case of water electrolysis powered with renewable electricity. Also, hydrogen filling station infrastructure is missing and would be very expensive to build up, different to the charging infrastructure needed for electric cars.

Life cycle assessment of electric car mobility according to the literature already available is complex. Most LCA data deal with the global warming potential. Since CO_2-equivalents emission during the operation is dominating the LCA in total, an electric car can already have ecoefficiency advantages when charged with grid electricity (500 to 600 g CO_2/kWh presumed). However, charging the electric car with renewable electricity (30 g CO_2/kWh) improves its LCA performance significantly. Ecoimpact of smaller BEV is also much better according to the high ecoimpact of the battery, which must increase parallel to the size of the car. Some LCA studies published so far modeled quite heavy BEV, which are additionally assumed to drive periodically at higher speeds, both inefficient for a BEV. In contrast, a small BEV like the electrified Smart presented here and moved locally as well as regionally only can have the most beneficial CO_2-impact. During an e-conversion of a used car, as shown with the Smart, life cycle CO_2 emissions can be reduced by more than 80%

compared to that known from ICEV. However, this is a first estimation under optimistic assumptions (e.g., battery lifetime), which is planned to be critically reviewed in a more detailed model later.

Life cycle impact of BEV in categories other than the global warming potential reveals a complex picture, although BEV demonstrates advantages over ICEV in most categories. Althaus [49] even concludes that 'carbon footprint is not sufficient as environmental performance indicator' here. One disadvantage of BEV is the acidification potential associated with the smelting processes of Cu, Ni, and Co since a lot of Cu and, in some battery types, Ni and Co also are essential elements of electrical components. Additionally, there are acidifying emissions of coal-fired power plants depending on the local value of this type of power production. However, to what extent the local nearly zero-emission advantage of electric cars is incorporated into LCA models is still a question. Toxic emissions like NOx and fine dust are today shifted to power plants through the use of BEV (quantified in [1]), where it is easier to limit and control them. The BEV advantage of a much lower noise emission, for example, is not appreciated so far (a guideline is in preparation).

REFERENCES

1. Helmers E: Bewertung der Umwelteffizienz moderner Autoantriebe – auf dem Weg vom Diesel-PKW-Boom zu Elektroautos. Umweltwiss Schadst Forsch. 2010, 22:564-578.
2. Angerer G, Marscheider-Weidemann F, Wendl M, Wietschel M: Lithium for future technologies - demand and supply with special emphasis on electric vehicles (in German). [http://www.elektromobilitaet.fraunhofer.de/Images/]
3. WHO: [http://www.who.int/mediacentre/factsheets/fs313/en/index.html] Air quality and health.
4. German advisory council on global change (WBGU): World in transition: [http://www.wbgu.de/ fileadmin/ templates/ dateien/ veroeffentlichungen/ hauptgutachten/ jg2011/ wbgu_jg2011_en.pdf] A social contract for sustainability.
5. Helmers E: Bitte wenden Sie jetzt – das Auto der Zukunft. Wiley VCH, Weinheim; 2009:204.
6. UNEP: [http://www.unep.org/transport/pcfv/PDF/HEV_Report.pdf] Hybrid electric vehicles. An overview of current technology and its application in developing and transitional countries.
7. Halderman JD, Martin T: Hybrid and Alternative fuel vehicles. Pearson Prentice Hall, New Jersey; 2008:448.

8. Nemry F, Brons M: [http://ftp.jrc.es/EURdoc/JRC58748_TN.pdf] Plug-in hybrid and battery electric vehicles.

9. Larminie J, Lowry J: Electric vehicle technology explained. Chichester. John Wiley & Sons, ; 2003:303.

10. Angerer G, Erdmann L, Marscheider-Weidemann F, Scharp M, Lüllmann A, Handke V, Marwede M: Raw materials for future technologies (in German). [http:// www.isi. fraunhofer.de/ isi-de/ n/ download/ publikationen/ Schlussbericht_lang_20090515_final. pdf]

11. Podewils C: Power for good times (in German). Photon 2010, 2010:36-46.

12. Amirault J, Chien J, Garg S, Gibbons D, Ross B, Tang M, Xing J, Sidhu I, Kaminsky P, Tenderich B: The electric vehicle battery landscape: opportunities and challenges. [http://cet.berkeley.edu/dl/BatteryBrief_final.pdf]

13. Gerschler B, Sauer DU: Batterien für Elektrofahrzeuge – Stand und Ausblick. Presentation, Berliner Energietage 2010.

14. Kloess M: The role of plug-in-hybrids as bridging technology towards pure electric cars: an economic assessment. [http://publik.tuwien.ac.at/files/PubDat_191393.pdf]

15. Kavalov B, Peteves SD: Impacts of the increasing automotive Diesel consumption in the EU. [http://www.eirc-foundation.eu/Publications/Energy/EUR%2021378%20 EN.pdf]

16. Forschungsstelle für Energiewirtschaft: Energieeffizienz alternativer Kraftstoffe aus Biomasse und Erdgas im Vergleich mit konventionellen Kraftstoffen. [http://www. wiba.de/download/wiba_p6.pdf]

17. Fritsche UR: Endenergiebezogene Gesamtemissionen für Treibhausgase aus fossilen Energieträgern unter Einbeziehung der Bereitstellungsketten. Kurzbericht im Auftrag des Bundesverbands der deutschen Gas- und Wasserwirtschaft e.V. Darmstadt: Öko-Institut 2007, :14.

18. European Union: Well-to-Wheels analysis of future fuels and associated automotive powertrains in the European context. [http:// / www.lowcvp.org.uk/ assets/ presentations/ CONCAWE_WTW_CNG_preliminary_Sept_03 .pdf]

19. Nylund NO, Laurikko J, Ikonen M: Pathways for natural gas into advanced vehicles. Part A: technology and fuels for new generation vehicles. [http:// / www.bctia.org/ files/ PDF/ transportation/ Natural_Gas_for_Advanced_Vehicles_-_2003.pdf]

20. International Energy Agency: Hydrogen production and distribution. IEA Energy Technology Essentials 2007, 5:1-4.

21. Kloess M: The road towards electric mobility – an energy economic view on hybrid- and electric vehicle. [http://publik.tuwien.ac.at/files/PubDat_183277.pdf]

22. EU: Well to wheel analysis of future automotive fuels and powertrains in the eurppean context. [http://www.co2star.eu/publications/Well_to_Tank_Report_EU.pdf]

23. Husain I: Electric and hybrid vehicles – design fundamentals. 2nd edition. CRC press, Boca Raton; 2011. 490 pp

24. Simpson A: Full-Cycle assessment of alternative fuels for light-duty road vehicles in Australia. [http:// / www.autospeed.com.au/ static/ downloads/ articles/ 110155_UP-DATED_Full-Cycle_Assessmen t_of_Alternative_Fuels_for_Light-Du ty_Road_Vehicles_in_Australia.pdf]

25. An F, Santini D: Assessing tank-to-wheel efficiencies of advanced technology vehicles. [http://cta.ornl.gov/TRBenergy/trb_documents/an_assessing_tank.pdf]

26. Pelz N: Alternative Kraftstoffe für Kraftfahrzeuge und ihre Möglichkeit zur CO2-Einsparung. In Im Spannungsfeld zwischen CO2-Einsparung und Abgasemissionsabsenkung. Expert-Verlag, Renningen; 2008:260.
27. Steiger W: Synthetic fuels – key for future power trains. In Which fuels for low CO-2engines. Edited by Pierre D. Technip, Paris; 2004:233. PubMed Abstract |
28. Bossel U: Does a hydrogen economy make sense? Proc IEEE 2006, 95(10):1826-1837.
29. Hacker F, Harthan R, Matthes F, Zimmer W: Environmental impacts and impact on the electricity market of a large scale introduction of electric cars in Europe. [http://acm.eionet.europa.eu/docs/ETCACC_TP_2009_4_electromobility.pdf]
30. Sullivan JJ, Baker RE, Boyer BA, Hammerle RH, Kenney TE, Muniz L, Wallington TJ: CO2emission benefit of diesel (versus gasoline) powered vehicles. Environ Sci Tech 2004, 38(12):3217-3223.
31. Albus C: WLTP-development – UNECE and the parallel EU process. [http://circa.europa.eu/ Public/ irc/ enterprise/ automotive/ library?l=/ cars_working_groups/ internal_emissions/ meeting_24012011/ presentation_germanypdf/ _EN_1.0_&a=d]
32. Arnold K, Dienst C, Lechtenböhmer S: Integrierte Treibhausgasbewertung der Prozessketten von Erdgas und industriellem Biomethan in Deutschland. Umweltwiss Schadst Forsch 2010, 22:135-152.
33. Majeau-Bettez G, Hawkins TR, Strømman AH: Life cycle environmental assessment of lithium-ion and nickel metal hydride batteries for plug-in hybrid and battery electric vehicles. Environ Sci Technol 2011, 45(10):4548-4554.
34. Auto Motor Sport 2011.
35. Matsunaga M, Fukushima T, Ojima K: Powertrain system of Honda FCX Clarity fuel cell vehicle. World Elec Vehicle J 2009, 3:1-10.
36. DWV: Wasserstoff: Der neue Energieträger. [http://www.dwv-info.de/publikationen/2009/etraeger3.pdf]
37. Simons A, Bauer C: Life cycle assessment of hydrogen use in passenger vehicles. [http://www.thelma-emobility.net/pdf/IAMF%202011/IAMF2011_Simons.pdf]
38. Linssen J, Grube T, Hoehlein B, Walbeck M: Full fuel cycles and market potentials of future passenger car propulsion systems. Int J Hydrogen Energ 2003, 28:735-741.
39. Joint Research Center: Well-to-wheels analysis of future automotive fuels and powertrains in the European context. [http://ies.jrc.ec.europa.eu/TWT]
40. Helms H, Pehnt M, Lambrecht U, Liebich A: Electric vehicle and plug-in hybrid energy efficiency and life cycle emissions. [http://www.ifeu.de/verkehrundumwelt/]
41. Pehnt M, Helms H, Lambrecht U, Lauwigi C, Liebich A: Umweltbewertung von Elektrofahrzeugen. Erste Ergebnisse einer umfassenden Ökobilanz. 14th Internationaler Kongress Elektronik im Kraftfahrzeug 2010. VDI-Berichte Elektronik im Kraftfahrzeug, Band Nr 2010, 2075:21-40.
42. Held M, Baumann M: Assessment of the environmental impacts of electric vehicle concepts. In Towards life cycle sustainable management. Edited by Houten Finkbeiner M. Springer Media, ; 2011:535-546.
43. Mitsubishi innovative electric vehicle: . [http://www.imiev.de/docs/iMiEV-daten.pdf]
44. Nissan Netherlands [http://www.nissan.nl]
45. ADAC-Motorwelt February

46. Eberhard M, Tarpenning M: The 21stcentury electric car. [http://www.fcinfo.jp/whitepaper/687.pdf]
47. About the ratings [http://www.fueleconomy.gov/feg/ratings2008.shtml]
48. German motor press
49. Althaus HJ: Comparative assertion of battery electric cars with various alternatives. [http://empa.ch/plugin/template/empa/*/109103]
50. International Energy Agency: CO2emissions from fuel combustion – highlights. [http://www.iea.org/co2highlights/co2highlights.pdf]
51. Yan X, Crookes RJ: Energy demand and emissions from road transportation vehicles in China. Prog Energy Combust Sci 2010, 36:651-676.
52. German Federal Environmental Agency
53. Opel Germany: . [http://www.opel.de]
54. Auto Motor Sport14.6.2011
55. AutoBild8.7.2011
56. Öko-Institut : Endenergiebezogene Gesamtemissionen für Treibhausgase aus fossilen Energieträgern unter Einbeziehung der Bereitstellungsketten. Darmstadt, ; 2007:14. PubMed Abstract | PubMed Central Full Text
57. www.adac.de 11/2011
58. AutoBild 10.6.2011
59. Höhlein B, Grube T: Kosten einer potenziellen Wasserstoffnutzung für E-Mobilität mit Brennstoffzellenantrieben. Energiewirtschaftliche Tagesfragen 2011, 61(6):62-66.
60. Wu Y, Wang MQ, Sharer PB, Rousseau A: Well-to-wheels results of energy use, greenhouse gas emissions and criteria air pollutant emissions of selected vehicle/fuel systems. SAE Transactions 2007, 115:210-222.
61. Notter DA, Gauch M, Widmer R, Wäger P, Stamp A, Zah R: H-J Althaus: Contribution of Li-ion Batteries to the Environmental Impact of electric vehicles. Environ Sci Tech 2010, 44:6550-6556.
62. Klöpffer W: B Grahl: Ökobilanz (LCA). Wiley-VCH, Weinheim; 2009. 426 pp
63. Ishihara K, Kihira N, Terada N, Iwahori T: Environmental burdens of large Li-ion batteries developed in a Japanese national project. In Central research institute of electric power industry, 202
64. Zackrisson M, Avellan L, Orlenius J: LCA of Li-ion batteries for plug-in hybrid electric vehicles – critical issues. J Clean Prod 2010, 18:1517.
65. Frischknecht R: LCA of driving electric cars and scope dependent LCA models. In 43rdLCA Forum. , Zürich; 2011.
66. Nissan Netherlands brochure 2011.
67. Frischknecht R, Flury K: Life cycle assessment of electric mobility: answers and challenges – Zurich, April 6, 2011. Int J Life Cycle Assess 2011, 16:691-695.
68. Althaus HJ, Notter D, Gauch M, Widmer W, Wäger P, Stamp A, Zah R: LCA of Li-ion batteries for electric mobility. [http://www.empa.ch/plugin/template/empa/*/104371]
69. Tao PC, Stefansson H, Harvey W, Saevarsdottir G: Potential use of geothermal energy sources for the production of Li-ion batteries. In Word Renewable Energy Congress. WREC, Linköping; 2011:8. 8–13

70. Samaras C, Meisterling K: Life cycle assessment of greenhouse gas emissions from plug-in hybrid vehicles: implications for policy. Environ Sci Tech 2008, 42:3170-3176.

71. Chanoine A: Comparative LCA of NiCd batteries used in cordless power tools vs. their alternatives NiMH and Li-ion batteries. [http://ec.europa.eu/environment/waste/batteries/pdf/biois_lca_18072011.pdf]

72. Canzler W, Knie A: Einfach aufladen – mit Elektromobilität in eine saubere Zukunft. Oekom Verlag, München; 2011.

CHAPTER 10

Investigated Cold Press Oil Extraction from Non-Edible Oilseeds for Future Bio-Jet Fuels Production

XIANHUI ZHAO, LIN WEI, JAMES JULSON, AND YINBIN HUANG

10.1 INTRODUCTION

As a result of concerning food vs. fuel debates, current biofuel development has focused on non-edible feedstock sources. Presently, a large number of non-edible vegetable oils are available, which do not compete with the food industry, so they can provide a source for bio-jet fuel production. In general, non-edible vegetable oil sources are classified into three categories: non-edible plant oils (e.g. *Camelina sativa, Jatropha curcas, Nicotiana tabacum* and *tamanu*), recycled oils derived from edible oilseed processing waste, and waste cooking oils. For example, genetically modified canola grown on margin lands has been identified as a sustainable bio-

Investigated Cold Press Oil Extraction from Non-Edible Oilseeds for Future Bio-Jet Fuels Production.
© Zhao X, Wei L, Julson J, and Huang Y. Journal of Sustainable Bioenergy Systems *4,4 (2014). doi:*
10.4236/jsbs.2014.44019. Licensed under Creative Commons Attribution License, http://creativecommons.org/licenses/by/3.0/.

fuel source because it doesn't occupy arable land. These margin lands are largely unproductive, or located in degraded forests and poverty-stricken areas. However, the canola plants are well adapted to arid and semi-arid conditions as they can grow on lands with low fertility and moisture, such as fallow lands, cultivators' field boundaries and old mining lands [1] [2]. Canola seeds contain about 40% oil and they are marketed worldwide [3] . Another non-edible vegetable oil producer is *Camelina sativa* seed, which is a non-edible oilseed that can be grown on marginal lands with a low input cost. One advantage of this oilseed is its resistance to blackleg, a disease that infects sunflowers, safflowers and many other crops. In Montana State of the United States, camelina is expected to have seed yield of 1800 to 2000 pounds per acre under dry land conditions [4] . Due to the widespread production, the camelina and canola seeds are considered important renewable energy sources.

Oilseeds produce oils that may be subsequently upgraded into saturated, unbranched and long-chain hydrocarbon fuels, which are suitable for bio-jet fuel production [5] . Using vegetable oils for bio-jet fuel production has potential advantages, such as high energy density, low moisture content and high-relative stability [6] [7] . Converting vegetable oils to bio-jet fuels includes three main steps. First, oil is extracted from vegetable oilseeds. Next, oil is upgraded into hydrocarbon fuel. Finally, hydrocarbon fuel is delivered to petroleum refinery for production of bio-jet fuel.

Vegetable oil extraction from oilseeds is a key step for bio-jet fuel production. Vegetable oil properties, specifically fatty acid profiles (FAPs) of oils, are dependent on oilseed species, oil extraction techniques and pro- cessing conditions. FAP is the quantitative composition of fatty acids in a vegetable oil. C16 and C18 fatty acids are the most common fatty acids in the vegetable oil. Fatty acids could affect the characteristics of vegetable oil, such as viscosity, oxidative stability, boiling point and combustion energy, so vegetable oil with good FAP is easily upgraded into a desired hydrocarbon fuel at high efficiency and low cost [8] [9] . Conventional methods of vegetable oil extraction include distillation, maceration, solvent extraction and cold press. The concentration of fatty acids in oil extracted from seeds using cold press method is similar to that using solvent extraction with petroleum-ether method, supercritical fluid extraction method and aqueous extraction method [2] . Each method has its own advantages and disadvantages. Using distillation technology consumes high energy and has slow output. The process of maceration is quite time-consuming. Solvent extraction is an expensive method with foam deposits in distillation. However, cold press has low capital cost with minimal labor. It is used in a small scale, which is suitable for oil extraction in rural areas. Also, it involves no organic solvent, and thus the product is chemically contaminant free. Additionally, the byproduct of vegetable oilseed meal has many potential applications. For example, the crude canola meal has been explored for use in dairy, poultry, floor covering, paper coatings, insulation and bio-composites [10] - [12] . Furthermore, recycling oilseed meals can increase oil extraction plant revenues, which could improve the economic viability of bio-jet fuel production [13].

Rombaut et al. have used screw pressing method to extract oils from grape seeds. They found that screw pressing was an efficient process for extracting grape seed oils with a high oil recovery [14] . Yetim et al. determined the fatty acid compositions of oils cold pressed from seven plant seeds in Turkey [10] . Biswas et al. used a catalytic cracking method to upgrade soybean oil. They found only liquid product with total absence of non-condensable gases when using a complex catalyst of Zr-Zr covalently bonded with alumina. Also, increasing reaction temperature could increase the alkane content with fewer fatty acids [15] . However, there are few related papers or reports about vegetable oil used as bio-jet fuel from purely experimental fuels to commercialization.

The goal of this study is to explore a sustainable pathway to produce bio-jet fuel from non-edible vegetable seeds. In the present work, oil extraction from non-edible camelina and canola seeds will be performed using the cold press method. The vegetable oils produced will be characterized. The effects of screw rotation speed frequency on the oil recovery and properties, such as moisture content, pH value, density, dynamic viscosity, elemental content and chemical composition will be discussed. In addition, a preliminary catalytic cracking test of camelina and canola oils produced at 15 Hz will be conducted to examine their potential for upgrading to hydrocarbon fuels.

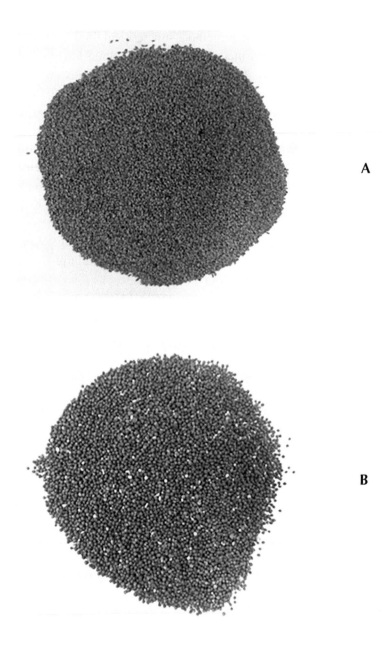

FIGURE 1: Oilseeds: (a) camelina (sativa) seeds; (b) canola (napus) seeds.

10.2 MATERIALS AND METHODS

10.2.1 MATERIAL AND DEVICE PREPARATION

Camelina (sativa) seeds were purchased from Hancock Seed Company, Dade, Florida, USA. Canola (*Brassica napus*) seeds were purchased from Agriculture and Agri-Food Canada, Ottawa, Ontario, Canada. These oilseeds, shown in Figure 1, were packed in bags and delivered to our biofuel laboratory. The bags were sealed and stored in the biofuel laboratory at room temperature for about four months before oil extraction started. There was no rancid smell or spoiling oilseed found during the oil extraction trials. These oilseeds had good quality prior to the tests. Both oilseed varieties were used directly without further treatment in this study. The M70 Oil Press was purchased from Oil Press Company, Eau Claire, Wisconsin, USA. It can produce about 70 gallons of oils from vegetable seeds and nuts each day running 24 hours. Table 1 lists the particle size, shape, bulk density and moisture content of camelina and canola seeds. The camelina seeds were quite small and had a rough surface. The length and width of seeds were measured based on the standard of ASTM D 4791. The size of seeds was tested using a proportional caliper device. The length means the maximum dimension of particles and the width refers to the maximum dimension in the plane perpendicular to the length. The moisture contents of seeds were determined by following the ASABE standards. The sample was dried in an oven at 105°C for 24 hours. Afterward, the dried sample was weighed to calculate the water content in the seeds [16] . Bulk density is an important physical characteristic of oilseeds that may affect the cold press processing. It is determined by following the ASABE standards. A cylindrical container with an inside diameter was used to measure the bulk density. The oilseed sample was poured into the container from a certain height to help with the free flow of particles. When the container was overfilled, a straight edge was stricken across the top of the container to remove the excess particles. Each measurement was repeated one time. The weight of the sample was measured. The height and inside diameter of the container was used to calculate its volume. Then, the bulk density was calculated by dividing the weight of sample over the volume of the container [17] . As shown in Table 1, camelina seeds had a slightly higher bulk density than canola seeds. The moisture of camelina seeds was less than that of canola seeds.

TABLE 1: The physical properties of flax and canola seeds.

Oilseeds	Length (mm)	Width (mm)	Shape	Bulk Density (g/mL)	Moisture (wt%)
Camelina	1.6 - 1.9	0.7 - 1.0	Oval	0.76 ± 0	4.92 ± 0.04
Canola	1.1 - 1.8	1.1 - 1.8	Tiny round	0.72 ± 0.02	5.36 ± 0.04

10.2.2 EXPERIMENTAL PROCEDURE

Oil extraction from camelina and canola seeds was carried out at different frequencies using the M70 Oil Press to discover the highest recovery of vegetable oils. The M70 Oil Press, as shown in Figure 2, mainly consists of a nozzle, a heat bond, meshes, a feeder, a motor, a frequency switch and a screw. The screw press is consisted of a horizontal screw and a vertical feeder. The Oil Press was operated to extract oils from vegetable oilseeds. The press barrel was preheated to 100°C, which was helpful for the oil extrusion. Before oilseeds were fed in through the vertical feeder, the screw was started at a setting frequency that was controlled by the Variable Frequency Drive (VFD) in order to avoid the blockage of oilseeds. Oilseeds were compressed and milled by the screw. Oils would be extruded out of the oilseeds and collected from small meshes below the screw. Friction created during the screw rotation provided heat for the oils and improved oil flow. Meanwhile, meals were removed through the nozzle in front of the screw. Based on the results of preliminary test runs, the Oil Press machine might not be able to work properly when the electrical frequency is greater than 25 Hz or less than 15 Hz. Therefore, the oilseeds were cold pressed at 15 Hz, 20 Hz, and 25 Hz respectively. The test runs were duplicated. Totally, twelve test runs were needed for the study.

FIGURE 2: M70 Oil Press.

After screw pressing the oilseeds, some oils remained in the meals. The meals were then ground using a mill (Thomas-Wiley Laboratory Mill, Model 4, Thomas Scientific , USA) and sieved through a 10 mesh screen. The meal flours were processed using a solvent extraction method to determine the residual oil content. An Accelerated Solvent Extractor (Dionex ASE 350, Thermo Scientific Company) was utilized to test the residual oil content in meals.

Camelina and canola oils cold pressed at 15 Hz were subjected to catalytic cracking in a fixed-bed reactor at 500°C at a liquid hourly space velocity (LHSV) of 1.0 h^{-1}. Nitrogen was used as the carrier gas in the reactor with a pressure of 1.38 × 105 Pa (20 psi). A preheater was used to vaporize oils for improved contact with the catalyst. ZSM-5 doped with 10 wt% of Zn was used as the catalyst, which was placed in a reactor. The reactor, fixed coaxially in a furnace, was a 508 mm long stainless steel tube with a 25.4 mm internal diameter. When oil vapors came in contact with the catalyst, cracking reactions took place. A condenser system setting at −10°C was used to cool the produced oil gases into liquid, considered as upgraded oils. Non-condensable gases, including H_2, CO, CO_2 and light hydrocarbons, were sampled for the composition analysis.

Upgraded camelina oils (UCMO) and upgraded canola oils (UCNO) were a mixture mainly containing hydrocarbons, acids and other oxygenates. They were distilled at 230°C in order to separate small molecules with lower boiling points. During distillation, small molecules with boiling point lower than 230°C became vapors. These vapors flowed into an overhead condenser system and were cooled back into liquid, considered as mixed hydrocarbons. Large molecules with boiling point higher than 230°C were too heavy to vaporize and they were remained as distillation residues in the distillation flask.

10.2.3 DATA COLLECTION

The obtained vegetable oils, upgraded oils, mixed hydrocarbons and distillation residues were characterized by testing their dynamic viscosity, pH value, moisture content, density, main chemical compositions, elemental content, heating value and yield.

Dynamic viscosity of samples was tested using a Visco Analyzer (REOLOGICA Instruments AB Company) at 20°C. The moisture content of samples was measured using a Karl Fischer Titrator V20 (Mettler Toledo Company) at 25°C, which is within ASTM E1064 standard. PH values were determined using a pH meter (Accumet BASIC AB15, Fisher Scientific) at 25°C and pH testing papers. The density of samples was measured by the ratio of mass to volume of the samples at room temperature [18] [19] .

The major chemical compositions of samples were analyzed by Gas Chromatography-Mass Spectrometry (GC-MS) (Agilent GC -7890A and MSD -5975C), that uses hydrogen as a carrier gas with a flow rate of 1.0 mL/min. The capillary columns are 30 m × 0.25 mm × 0.25 μm DB-5MS. The samples were prepared following the derivatization procedure. About 30 mg sample was mixed with 4 mL hexane. The mixture was shaken for 2 min, and then 2 mL BF_3-methanol, 12% w/w, and 2 mL CH_3OH were added into the mixture. The mixture was heated at 60°C for 10 min to carry out the simultaneous hydrolysis and methylation. Then, the mixture was cooled down to room temperature and 1 mL distilled water and 2 mL hexane were added in order to remove the excess reagents. The organic phase of the mixture was separated by centrifugation at 2500 g and 25°C for 10 min. The organic layer was carefully removed and dried over anhydrous sodium sulfate. Finally, the dry organic phase was injected into the GC-MS equipment. The GC-MS test parameters were as follows. One μL of the dry organic phase was introduced through the injection port operated in a splitless mode at 260°C. The original column temperature was 175°C and the holding time was 6 min. Then, the column temperature became 260°C after 5.67 min at a rate of 15°C/min. The holding time at 260°C was 15.68 min. The splitless time was 30 s and the total run time was 27.35 min. The major components were identified through the NIST Mass Spectral library [12] [20] .

Carbon (C), hydrogen (H), nitrogen (N), and oxygen (O) content analysis of samples was determined using a CE-440 Elemental Analyzer (Exeter Analytical. Inc.) according to ASTM D4057 standard. Acetanilide was used for calibration. For CHN content analysis, the combustion and reduction temperatures were 980°C and 650°C, respectively. The oxygen and helium pressure were 1.51×10^5 Pa (22 psi) and 1.17×10^5 Pa (17 psi), respectively. The fill time was between 20 and 50 s. The combustion and purge time were both 20 s. Samples were sealed in a tin capsule and placed in a nickel sleeve. For O content analysis, the combustion and reduction temperature were 960°C and 770°C, respectively. Only helium was used as the carried gas at 1.17×10^5 Pa (17 psi). The combustion and purge time were 40 s and 50 s, respectively. The sample was sealed in a silver capsule and placed inside a nickel sleeve [21].

The heating value was tested using a C 2000 Calorimeter System (IKA-Works, Inc.) according to the temperature change of water inside the measuring cell, based on ASTM D 4809 standard. About 0.5 g sample was added in a crucible and placed in a bomb. The sample was ignited with a cotton twist.

The Accelerated Solvent Extractor was used to determine the residual oil content inside the vegetable oilseed meals. About 12 g meal flours were used in this test and hexane was used as the solvent. The oven temperature was 105°C and the static time was 10 min. The rinse volume was 50% and the purge time was 60 s. The meal flours were mixed with hexane and the oil inside was dissolved. After the oil was dissolved, the mixture of oil and hexane was separated from the solid materials of the meal flours. Then, the oil was separated from hexane by using a distillation system and the oil content remaining in meals was calculated. The oil extraction was completed with three static cycles. The tests for oil content in each residual meal sample were carried out three times [22] [23].

The oil content of oilseeds depends on many factors such as variety, fertilization, growth environment and agricultural production technologies [24]. In this study, the oil content of oilseeds (camelina and canola seeds) was defined as the sum of the oil yield using cold press method and the oil content of residual meals. The oil recovery of oilseeds was defined as the oil yield using cold press method divided by the oil content of oilseeds.

The yields of products were defined by the following equations:

Yield of upgraded oils = (mass of upgraded oils/mass of oils feed) × 100%   (1)

Yield of mixed hydrocarbons =
    (mass of mixed hydrocarbons/mass of upgraded oils feed) × 100%   (2)

Yield of distillation residues =
  (mass of distillation residues/mass of upgraded oils feed) × 100%   (3)

10.2.4 STATISTICAL ANALYSIS

In this study, all treatments were conducted in duplicate. The determination of the mean and standard deviation of each parameter was carried out using Microsoft Excel 2013 (Microsoft Corp., Redmond, WA).

10.3 RESULTS AND DISCUSSION

10.3.1 CHEMICAL COMPOSITIONS

Table 2 shows the major chemical compounds in the camelina and canola oils produced at three different frequencies using the cold press method. The GC-MS analysis of all camelina oils showed that the carbon distribution lied between C14 and C22. Camelina oils, produced at 25 Hz, 20 Hz, and 15 Hz, all contained 9,12,15-Octa- decatrienoic acid, (Z,Z,Z)-, occupying 82.4%, 84.5%, and 78.7%, respectively. Also, all camelina oils contained cis-11-eicosenoic acid, occupying 5.00% - 14.5%. However, in camelina oils produced at 15 Hz, the total content of 9,12-Octadecadienoic acid, (E,E)- (C18), tridecanoic acid, 12-methyl-(C14), 8,11,14-Eicosatrienoic acid (C20) and oleic acid (C18) was 9.10%. It indicated that the main compositions in camelina oils produced at three frequencies were the same, while other minor compositions were different. The GC-MS analysis of all canola oils showed that the carbon distribution lied between C9 and C20. For canola oils produced at 25 Hz, 20 Hz, and 15 Hz, they all contained the same composition of oleic acid, occupying 98.6%, 92.5%, and 97.2%, respectively. It also indicated that the main composition in canola oils produced at three frequencies was the same, while other minor compositions were different.

TABLE 2: Main compositions in camelina and canola oils produced at 25 Hz, 20 Hz and 15 Hz.

Camelina		Canola	
Component	Area (%)	Component	Area (%)
25 Hz			
Hexadecanoic acid ($C_{16}H_{32}O_2$)	0.23	Hexadecanoic acid ($C_{16}H_{32}O_2$)	0.18
9,12-Octadecadienoic acid (Z,Z)- ($C_{18}H_{32}O_2$)	0.15	Oleic acid ($C_{18}H_{34}O_2$)	98.6
9,12,15-Octadecatrienoic acid, (Z,Z,Z)- ($C_{18}H_{30}O_2$)	82.4	Octadecanoic acid ($C_{18}H_{36}O_2$)	1.05
Octadecanoic acid ($C_{18}H_{36}O_2$)	1.11	9-Octadecenamide, (Z)- ($C_{18}H_{35}NO$)	0.20
cis-11-eicosenoic acid ($C_{20}H_{38}O_2$)	14.5	Methyl 18-methylnonadecanoate ($C_{21}H_{42}O_2$)	0.22
9-Octadecenamide ($C_{18}H_{35}NO$)	0.12	13-Docosenoic acid, (Z)- ($C_{22}H_{42}O_2$)	1.27
20 Hz			
Pentadecanoic acid, 14-methyl- ($C_{16}H_{32}O_2$)	0.14	Hexadecanoic acid ($C_{16}H_{32}O_2$)	0.89
9,12,15-Octadecatrienoic acid, (Z,Z,Z)- ($C_{18}H_{30}O_2$)	84.5	Oleic acid ($C_{18}H_{34}O_2$)	92.5
Octadecanoic acid ($C_{18}H_{36}O_2$)	0.80	Octadecanoic acid ($C_{18}H_{36}O_2$)	5.61
cis-11-eicosenoic acid ($C_{20}H_{38}O_2$)	13.7	cis-11-Eicosenoic acid ($C_{20}H_{38}O_2$)	0.62
Methyl 18-methylnonadecanoate ($C_{21}H_{42}O_2$)	0.23	9-Octadecenamide, (Z)- ($C_{18}H_{35}NO$)	0.44
13-Docosenoic acid ($C_{22}H_{42}O_2$)	0.62		
15 Hz			
Tridecanoic acid, 12-methyl- ($C_{14}H_{28}O_2$)	2.96	Hexadecanoic acid ($C_{16}H_{32}O_2$)	0.40
9,12-Octadecadienoic acid, (E,E)- ($C_{18}H_{32}O_2$)	3.06	9,15-Octadecadienoic acid ($C_{18}H_{32}O_2$)	0.71
9,12,15-Octadecatrienoic acid, (Z,Z,Z)- ($C_{18}H_{30}O_2$)	78.7	Oleic acid ($C_{18}H_{34}O_2$)	97.2
Oleic acid ($C_{18}H_{34}O_2$)	1.47	6-Nonynoic acid ($C_9H_{14}O_2$)	0.28
cis-11-eicosenoic acid ($C_{20}H_{38}O_2$)	5.00	8,11,14-Eicosatrienoic acid ($C_{20}H_{34}O_2$)	1.61

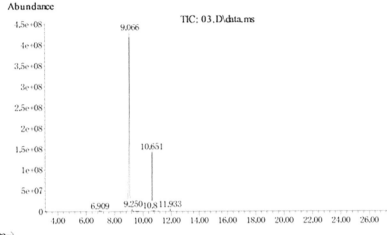

A

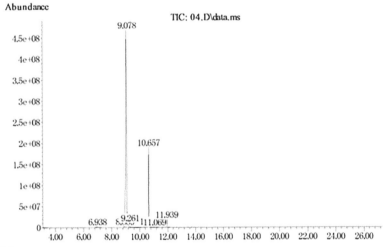

B

FIGURE 3: GC-MS chromatograms of oil samples: (a) camelina oil produced at 25 Hz; (b) camelina oil produced at 20 Hz; (c) camelina oil produced at 15 Hz; (d) canola oil produced at 25 Hz; (e) canola oil produced at 20 Hz; (f) canola oil produced at 15 Hz.

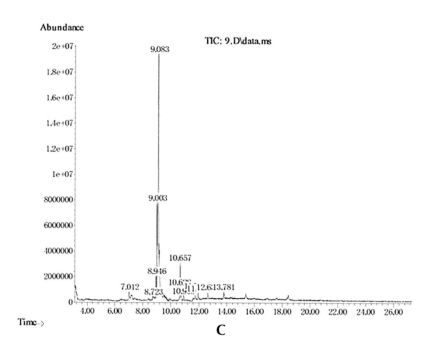

C

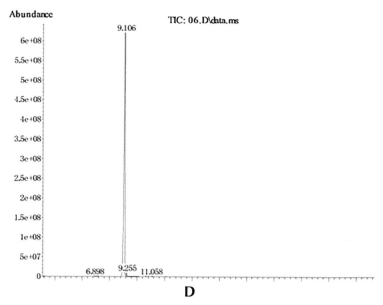

D

FIGURE 3: *Cont*

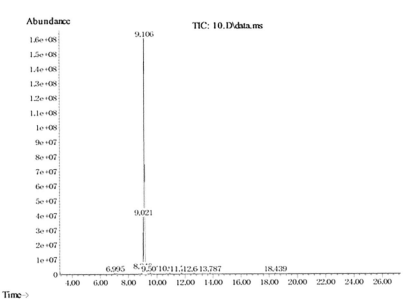

E

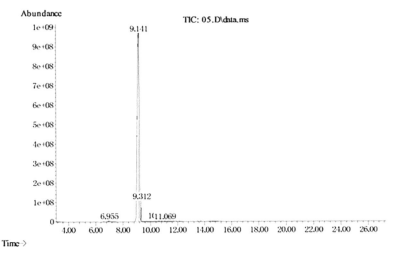

F

FIGURE 3: *Cont*

For both camelina and canola oils, they mainly contained fatty acids. Fatty acid profile is one of the important factors that affect the upgrading of vegetable oils for future jet fuel production. Furthermore, unsaturated fatty acids occupied most of the fatty acids in camelina and canola oils. These unsaturated fatty acids varied in the extent of unsaturation and in the carbon chain length. However, the main unsaturated fatty acid in both camelina and canola oils contained the same carbon chain length, C18.

Figure 3 represents the GC-MS chromatograms that resulted from camelina and canola oils cold pressed at three different frequencies. The GC-MS chromatograms showed that all camelina oils had two apparent high peaks. The main chemical composition in all camelina oils was 9,12,15-Octadecatrienoic acid, (Z,Z,Z)-, which was assigned at the highest peak. The 9,12,15-Octadecatrienoic acid, (Z,Z,Z)- was assigned at the peak eluting

If 9.078 min, 9.066 min, and 9.083 min for camelina oils produced at 25 Hz, 20 Hz, and 15 Hz, respectively. The second highest peak was assigned as cis-11-eicosenoic acid at about 10.65 min. Similarly, all canola oils had a same high peak, which was assigned as oleic acid. For canola oils produced at 25 Hz, 20 Hz and 15 Hz, oleic acid was assigned at the peak eluting of 9.106 min, 9.141 min, and 9.106 min, respectively. The properties of the fatty acids, such as branching of the chain, chain length and degree of unsaturation, could influence the bio-fuel quality. Both 9,12,15-Octadecatrienoic acid, (Z,Z,Z)- and oleic acid were found suitable for hydrocarbon fuel production [25] - [27] . For camelina oils produced at 15 Hz, there appeared some small peaks after 12.0 min. For canola oils produced at 15 Hz, there also appeared some small peaks after 11.1 min. This might be due to the lower frequency that caused a slower extraction of oilseeds.

10.3.2 ELEMENTAL CONTENT

Table 3 lists the CHNO content of camelina and canola oils produced at three frequencies. Carbon was a main component in all oil samples. The nitrogen content of all oil samples was lower than 0.8%, which indicates a low level of phospholipid. For all camelina oils produced at the three frequencies, the carbon content was the same. Also, their H/C molar ratios were between 1.81 and 1.82. Canola oils had a lower carbon content and higher hydrogen content than camelina oils. The H/C molar ratios of canola oils were between 1.86 and 1.87. It means the frequency during the cold press processing of camelina and canola seeds had less effect on the CH content. The H/C molar ratios of both camelina and canola oils were lower than that of petroleum product, which was 2.0.

Vegetable oils contain oxygen atoms in the structure of hydrocarbons. The oxygen content influences the specific energy and the combustion [28] . During vegetable oil upgrading, removing oxygen is a significant aim. From Table 3, the oxygen contents of camelina oils cold pressed at different frequencies were between 11.7% and 12.6%, indicating that they had no big difference. Similarly, the frequency during the cold press processing of oilseeds had less effect on the oxygen content of canola oils. For bio-jet fuel, the oxygen content is below 1.0% [29] . Therefore, the oils need to be further treated in order to reduce the oxygen content.

10.3.3 PHYSICAL PROPERTIES

The densities of camelina oils produced at different frequencies, shown in Table 4, were quite similar. For canola oils, the density values showed a minor difference. Camelina oil produced at 20 Hz had the highest density, which was 0.90 g /mL. The density of oil is related to its chemical structure and composition. During the cold press processing of canola oilseeds, the temperatures below the heat bond were slightly different. For canola oils produced at 20 Hz, the temperature was between 100.8°C and 103.7°C. However, the temperatures were within the range of 94.9°C - 98.2°C at 15 Hz and 97.1°C - 98.8°C at 25 Hz, respectively. The slight difference in temperatures possibly occurred due to the different chemical structure and composition of oils, thus leading to the different densities. However, the densities of both camelina and canola oils were between 0.87 and 0.90 g / mL, which were higher than that of jet fuel (0.81 g /mL) [30] -[32] .

The pH values of camelina and canola oils produced at different frequencies are shown in Table 4. Acid value is usually used to quantify the amount of acids in samples. It is a more standardized measurement than pH value. However, only the pH values were measured in this study to qualitative the amount of organic acid in samples due to the limitation of current equipment conditions. In the future research, the acid value of samples will be determined. All the oils produced from camelina and canola oilseeds showed a mild acidity. The pH values of camelina oil produced at 15 Hz and 25 Hz were the same, while the pH value of camelina oil produced at 20 Hz was slightly different from them. For canola oils produced at different frequencies, the pH values were not very different and they varied between 3.75 and 4.09. At a low temperature range of 94.9°C - 108.9°C during the cold press machine process, mechanical aspects may have less effect on the pH. The acidity of vegetable oils reflects the content of free fatty acids, which are produced from chemical or enzymatic (lipolytic enzymes) reactions that cause the split of triglycerides, monoglycerides and diglycerides [2] . Since camelina oils had a different content of fatty acids from canola oils, there was a big difference between the pH values of these two types of oils.

TABLE 3: The CHNO content of oil samples.

Oils	Frequency (Hz)	C (wt%)	H (wt%)	N (wt%)	O (wt%)
Camelina	25	78.4 ± 0.05	11.9 ± 0.01	0.40 ± 0.25	12.6 ± 0.17
Camelina	20	78.4 ± 0.04	11.9 ± 0.01	0.29 ± 0 .07	12.4 ± 0.78
Camelina	15	78.4 ± 0.01	11.8 ± 0.04	0.71 ± 0.21	11.7 ± 0.76
Canola	25	77.7 ± 0.07	12.1 ± 0.01	0.57 ± 0.07	12.1 ± 0.79
Canola	20	77.4 ± 0.06	12.0 ± 0.04	0.55 ± 0.42	11.2 ± 0.47
Canola	15	77.6 ± 0.59	12.1 ± 0.16	0.51 ± 0.35	12.5 ± 0.62

The moisture content of oil samples at different frequencies is represented in Table 4. The free or bonded water in the oils results in the formation of free fatty acids, which can corrode the engine and fuel storage tank [26] [27] . For both camelina and canola oils, their water contents were very low, equal to or below 0.11%. The moisture content of Jet A/Jet A-1 was less than 0.1%, indicating that vegetable oils have potential for future jet fuel production [29] [33] . For both camelina and canola oils, the frequency had a minor effect on the moisture content of oils. Small amounts of water left in oils could help improve their mobility. The water in oils may come from the water within camelina seeds and canola seeds. Camelina oils contained less water than canola oils, which may result from the lower water content in camelina seeds compared to canola seeds. Most of the water in oilseeds had been evaporated during the cold press processing due to the temperature varying between 94.9°C and 108.9°C. Also, some water may be left in the extruded meals. The low moisture content of oils may be connected with their high viscosity, which was stated below.

TABLE 4: The physical properties of oil samples.

Oils	Frequency (Hz)	Density (g/mL)	pH Value	Moisture (%)	Viscosity (cP)	Heating Value (MJ/Kg)
Camelina	25	0.88 ± 0.04	5.14 ± 0.13	0.06 ± 0.01	58.9 ± 0.06	39.7 ± 0.06
Camelina	20	0.89 ± 0.02	5.44 ± 0.38	0.08 ± 0.01	59.4 ± 0.11	39.4 ± 0
Camelina	15	0.89 ± 0.01	5.14 ± 0.25	0.06 ± 0	59.7 ± 0.04	39.6 ± 0.05
Canola	25	0.87 ± 0.02	3.80 ± 0.63	0.11 ± 0	76.2 ± 0.09	39.7 ± 0.02
Canola	20	0.90 ± 0.01	3.75 ± 0.39	0.11 ± 0	76.5 ± 0.06	39.7 ± 0.06
Canola	15	0.89 ± 0.01	4.09 ± 0.86	0.10 ± 0	77.9 ± 0.04	39.7 ± 0

The dynamic viscosity of oils is shown in Table 4. Viscosity is one of the determining factors of fuel quality and use. It may significantly influence the performance of the pump and fuel injector in engines. Viscosity is affected largely by the chemical structure of oils, such as the fatty acid profile, triglyceride composition, chain length, chain branching, degree of saturation, molecular configuration (cis-trans, conjugation), oxidation presence and degradation products. Camelina oils produced at different frequencies had similar viscosities. Also, the viscosities for canola oils had minor differences. However, the viscosity of camelina oils was lower than that of canola oils. The main chemical composition in camelina and canola oils was 9,12,15-Octadecatrienoic acid, (Z,Z,Z)- ($C_{18}H_{30}O_2$) and oleic acid ($C_{18}H_{34}O_2$), respectively. The smaller chemical molecule size in camelina oils may result in the lower viscosity. Compared to the dynamic viscosity of

Jet A/Jet A-1, 2 cP measured at 25°C, the viscosities of camelina and canola oils were much higher [29] . Camelina and canola oils contained large chemical molecules and complicated molecular structures, which caused their high viscosity. It indicates that the direct use of vegetable oils as fuel is impractical and the oils need to be upgraded for the future bio-jet fuel use.

The heating values of oils are shown in Table 4. All of the heating values of camelina and canola oils had no significant difference. Perhaps the similar oxygen contents of them led to the similar heating values. The net heat value of Jet A/Jet A-1 was 43.2 MJ/Kg, indicating the vegetable oils have to be upgraded before being used as bio-fuels in the future [30] .

10.3.4 OIL RECOVERY

Figure 4 shows the oil recovery of camelina and canola seeds cold pressed at different frequencies. The highest oil recovery of 88.2% was achieved when camelina seeds were cold pressed at 15 Hz. The oil recovery of camelina seeds cold pressed at 20 Hz and 25 Hz was 88.0% and 86.7%, respectively. For canola seeds cold pressed at 15 Hz, 20 Hz and 25 Hz, the oil recovery was 84.1%, 83.6% and 83.3%, respectively. It is obvious that for both camelina and canola seeds, the oil recovery increased with the decrease of frequency. A lower frequency led to a slower extraction of oilseeds, which might increase the oil yield, thus resulting in a higher oil recovery. However, a higher oil recovery was obtained for camelina seeds, compared to canola seeds. The processing temperature was within a range of 99.9°C - 108.9°C during cold press processing of camelina seeds. For canola oils produced at three frequencies, the processing temperature was between 94.9°C and 103.7°C. The minor difference of the processing temperature range might be contributed to the slightly different shell hardness of camelina and canola seeds. Under this situation, the oil recovery of camelina and canola seeds became different. During cold press processing of vegetable oilseeds, meal was extracted as a co-product. For cold press processing of camelina seeds at these three frequencies, the meal yield was between 61.8% and 62.7%. For canola seeds extracted at these three frequencies, the meal yield was in a range of 61.1% - 61.5%. The oilseed meals with high yields have potential as food for animals, such as fish and cattle.

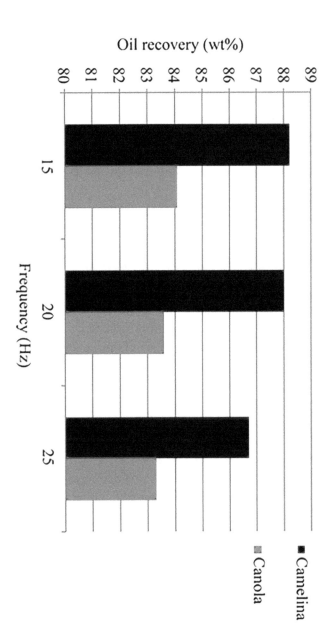

FIGURE 4: Oil recovery (wt%) of camelina and canola seeds.

10.3.5 OIL CONTENT OF RESIDUAL MEALS

Table 5 lists the oil content of residual meal extracted from camelina and canola seeds at 15 Hz. The solvent extraction method was capable of recovering about 99% oil contained in the oilseeds [34] . In this study, the oil content of residual meals was defined as the measured oil content of residual meals using solvent extraction method divided by 99%. The oil content of residual camelina meal was lower than that of residual canola meal. This is maybe because the hulls of camelina and canola seeds were different, thus resulting in different oil contents in the residual meals.

10.3.6 PRELIMINARY TESTS OF OIL UPGRADING

10.3.6.1 PHYSIOCHEMICAL CHARACTERIZATION

This is a preliminary study investigating the upgrading of camelina and canola oils, which were cold pressed at 15 Hz. In the present study, it is aimed to find out an effective cold press for oil extraction from camelina and canola seeds. Comparing the oil extraction from these two feed stocks using the cold press machine at the three different frequencies, the highest oil recovery was obtained for both feed stocks when the cold press ran at 15 Hz. In addition, the cold press running frequencies had slight influence on the oil properties. Therefore, the frequency of 15 Hz during the cold press processing of oilseeds was identified as the optimal processing parameter. Table 6 shows the main chemical compositions of products after catalytic cracking of camelina and canola oils at 500°C. Both upgraded camelina oils and upgraded canola oils mainly contained fatty acids, hydrocarbons, and esters. The total fatty acids content of upgraded camelina oils and upgraded canola oils was 59.84% and 76.85%, respectively. This meant that a certain amount of oxygen was removed during the catalytic cracking of camelina and canola oils. After distillation, the main component in both camelina mixed hydrocarbons (CMMH) and canola mixed hydrocarbons (CNMH) was hydrocarbons, which occupied between 70.53% and 74.67%. As shown in Figure 5, for camelina oil upgrading, apparent decreases in peak areas and peak heights (after 12.225 min) were observed in the GC-MS profiles of camelina mixed hydrocarbons than upgraded camelinaoils. Similarly, for canola oil upgrading, apparent decreases in peak areas and peak heights (after 12.133 min) were observed in the GC-MS profiles of canola mixed hydrocarbons than upgraded canola oils. It indicated that low-molecule weight compounds were separated and became mixed hydrocarbons after the distillation treatment [30] . For canola distillation residues (CNDR), they contained similar main components as upgraded canola oils. More obviously, camelina distillation residues (CMDR) contained similar main components as upgraded camelina oils. However, some tars might be generated during the upgrading process, which then remained in the distillation residues after the distillation treatment. This might cause a much higher viscosity of distillation residues than upgraded oils, as shown in Table 7.

TABLE 5: The oil content of residual meals.

Oilseeds	Oil Content of Residual Meal (%)
Camelina	4.90 ± 1.55
Canola	7.07 ± 1.21

Table 7 shows the physical properties of products during the upgrading of camelina and canola oils, which were cold pressed at 15 Hz. Raw camelina and canola oils were upgraded in order to improve their undesirable properties. After upgrading, there was a small decrease in density. The larger molecules were broken into smaller ones during the catalytic cracking process, which may result in the lower density. The viscosities of upgraded oils decreased greatly than raw oils. The lower viscosity made the oil easier for subsequent operating and pumping. In addition, the heating values of upgraded oils increased. However, the moisture content increased due to the water production during the upgrading process. After a treatment of distillation, the densities and viscosities of mixed hydrocarbons decreased further. Also, the moisture content of mixed hydrocarbons became lower than upgraded oils. These three properties were considered as good signs. The density, viscosity and moisture content of camelina mixed hydrocarbons have met the requirements of bio-jet fuel. The viscosity and moisture content of canola mixed hydrocarbons have met the requirements of bio-jet fuel. It indicates that the catalytic cracking method is effective to convert vegetable oils to mixed hydrocarbons, which have the potential for future bio-jet fuel production. For distillation residues, they had lower pH values and moisture content, and higher densities and viscosities than upgraded oils. In the future, the distillation residues could be treated further, such as a second catalytic cracking, for bio-fuels production.

10.3.6.2 THE YIELDS

Figure 6 shows the yields of upgraded camelina oils, camelina mixed hydrocarbons, camelina distillation residues, upgraded canola oils, canola mixed hydrocarbons, and canola distillation residues during the catalytic cracking and distillation process. The yields of upgraded camelina oils and upgraded canola oils cracked at 500°C were both higher than 60%. After distillation, the yields of camelina mixed hydrocarbons and canola mixed hydrocarbons were 21.9% and 26.7%, respectively. It indicated the distillation efficiency might be affected by the chemical compositions in upgraded oils.

10.3.6.3 GC ANALYSIS

Atypical GC analysis for the non-condensable gases produced from catalytically cracking camelina and canola oils at 500°C is shown in Figure 7. For TCD signal analysis, there were noise peaks between H_2 peak and CO_2 peak. These noise peaks were not drawn in the TCD signal figures in this study. Inside the ZSM-5-Zn-10% catalyst, there were several reactions occurring, such as dehydration, decarbonylation, decarboxylation, and isomerization. These reactions can convert carbon and hydrogen into olefins and aromatics; and they can remove oxygen as CO, CO_2, and H_2O [35] .

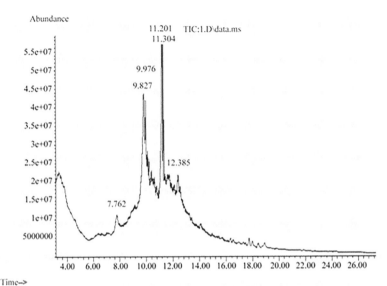

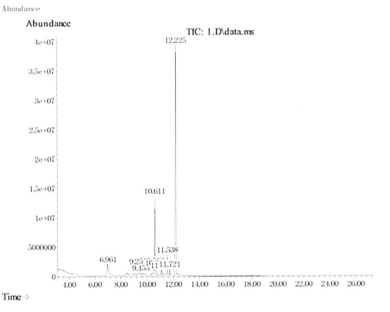

FIGURE 5: GC-MS chromatograms of products from catalytic cracking of camelina and canola oils at 500°C: (a) UCMO*; (b) CMMH; (c) CMDR; (d) UCNO; (e) CNMH; (f) CNDR. *UCMO: upgraded camelina oils; CMMH: camelina mixed hydrocarbons; CMDR: camelina distillation residues; UCNO: upgraded canola oils; CNMH: canola mixed hydrocarbons; CNDR: canola distillation residues.

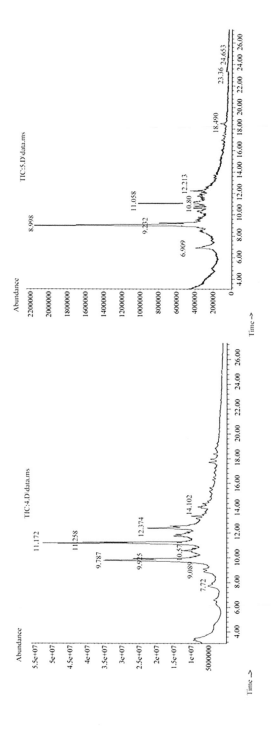

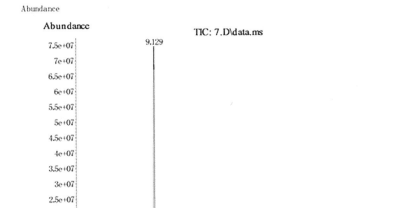

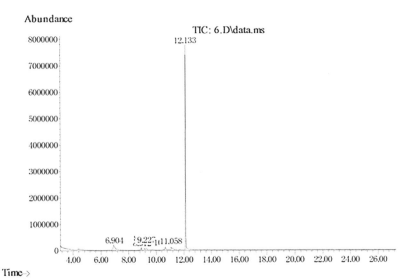

FIGURE 5: *Cont.*

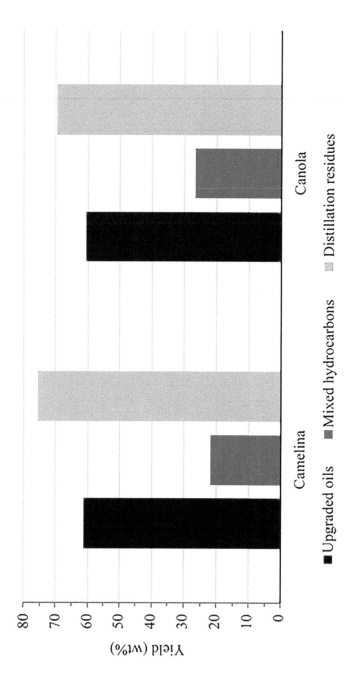

FIGURE 6: The yields of UCMO, CMMH, CMDR, UCNO, CNMH and CNDR.

During the catalytic cracking of camelina and canola oils, the non-condensable gases contained CH_4, C_2H_6, C_2H_4, C_3H_8, C_3H_6, C_4H_{10}, C_5H_{12}, H_2, CO_2 and CO. The production of CH_4 indicated that ZSM-5-Zn-10% was able to crack and convert fatty acids into the smallest fraction of hydrocarbons. The production of H_2 resulted from the dehydrogenation reaction. However, some H_2 was attracted to unstable species to produce more stable products. The production of CO and CO_2 were contributed to some light olefins, such as C_3H_6 and C_2H_4.

During the catalytic cracking of camelina oils, the total organic compositions, C1 to C5, occupied an area content of 12.14%. The area contents of H_2, CO_2, and CO were 5.25%. During the catalytic cracking of canola oils, the total organic compositions occupied an area content of 21.50%. The area contents of H_2, CO_2, and CO were 11.76%. It indicated that there were more decarboxylation and decarbonylation occurring during the catalytic cracking of canola oils at 500°C. The H_2 and CO generated during the catalytic cracking of camelina and canola oils have the potential for future syngas application.

10.4 CONCLUSIONS

The highest oil recovery for camelina and canola seeds using the cold press method is 88.2% and 84.1%, respectively. Also, more than 60% of meals are obtained during the cold press processing of these two seeds. The meals could be analyzed and treated in the future as food for animals. The density, pH value, moisture, heating value and oxygen content of both camelina and canola oils, produced at different frequencies, have no significant difference because of the relative low processing temperature during screw pressing. During the cold press of camelina and canola oilseeds, the frequency of 15 Hz is identified as the optimal processing parameter.

Camelina oils produced at three different frequencies all contain 9,12,15-Octadecatrienoic acid, (Z,Z,Z)-, which is the main component. All canola oils produced at the three different frequencies also contain the same main component, oleic acid. However, other minor compositions in camelina oils and canola oils produced at the three different frequencies are different.

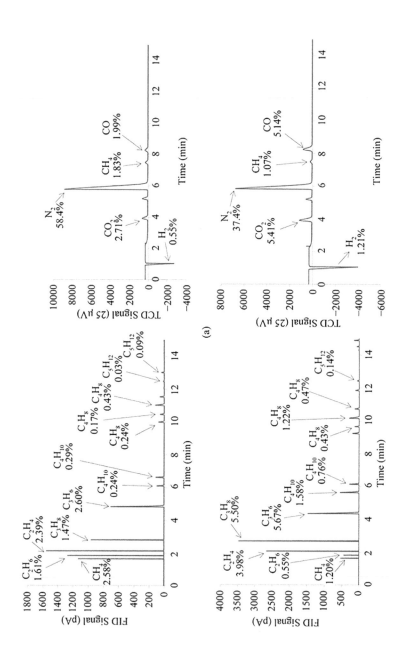

FIGURE 7: A typical GC analysis for non-condensable gases produced at 500°C from catalytic cracking of (a) camelina oils and (b) canola oils.

The undesired properties of camelina and canola oils are improved after the catalytic cracking to some extent. After catalytic cracking, the density and viscosity of upgraded oils decrease, and the heating value increases. After the distillation treatment, the densities and viscosities of mixed hydrocarbons decrease further. The mixed hydrocarbons, which have met the moisture and viscosity requirements of bio-jet fuel, have the potential for future bio-jet fuel production. However, more work needs to be done for the effective upgrading of non-edible vegetable oils for future bio-jet fuel production. For example, several important parameters, such as catalyst activity, reaction temperature and liquid hourly space velocity during the catalytic cracking of vegetable oils, could be considered. In addition, recycling catalysts that have been used during the oil upgrading could reduce the processing cost of converting vegetable oils to bio-jet fuels.

REFERENCES

1. Miraculas, G.A., Bose, N. and Raj, R.E. (2014) Optimization of Process Parameters for Biodiesel Extraction from Tamanu Oil Using Design of Experiments. Journal of Renewable and Sustainable Energy, 6, Article ID: 033120. http://dx.doi.org/10.1063/1.4880216

2. Zhao, X., Wei, L. and Julson, J. (2014) First Stage of Bio-Jet Fuel Production: Non-Food Sunflower Oil Extraction Using Cold Press Method. AIMS Energy, 2, 193-209. http://dx.doi.org/10.3934/energy.2014.02.193

3. Macrae, I., Oelke, E.A., Hutchison, W.D. and Nelson , J.J. (2000) Crop Profile for Canola in Minnesota. Minnesota Pesticide Impact Assessment Program (PIAP). http://www.ipmcenters.org/cropprofiles/docs/mncanola.pdf

4. Ehrensing, D.T. and Guy , S.O. (2008) Camelina. Extension Service, Oregon State University, 1-7. http://extension.oregonstate.edu/gilliam/sites/default/files/Camelina_em8953-e.pdf

5. Maher, K.D. and Bressler, D.C. (2007) Pyrolysis of Triglyceride Materials for the Production of Renewable Fuels and Chemicals. Bioresource Technology, 98, 2351-2368. http://dx.doi.org/10.1016/j.biortech.2006.10.025

6. Atabani, A.E., Silitonga, A.S., Ong, H.C., Mahlia, T.M.I., Masjuki, H.H., Badruddin, I.A. and Fayaz, H. (2013) Non- Edible Vegetable Oils: A Critical Evaluation of Oil Extraction, Fatty Acid Compositions, Biodiesel Production, Characteristics, Engine Performance and Emissions Production. Renewable and Sustainable Energy Reviews, 18, 211-245. http://dx.doi.org/10.1016/j.rser.2012.10.013

7. Cheng, J., Li, T., Huang, R., Zhou, J. and Cen, K. (2014) Optimizing Catalysis Conditions to Decrease Aromatic Hydrocarbons and Increase Alkanes for Improving Jet

Biofuel quality. Bioresource Technology, 158, 378-382. http://dx.doi.org/10.1016/j.
biortech.2014.02.112

8.   Demirbas, A. (2003) Fuel Conversional Aspects of Palm Oil and Sunflower Oil.
     Energy Sources, 25, 457-466. http://dx.doi.org/10.1080/00908310390142451

9.   Ma, Z., Wei , L., Qu, W., Julson, J., Zhu, Q. and Wang, X. (2013) The Effect of Sup-
     port on the Catalytic Performance for Bio-Oil Upgrading. Advanced Materials Re-
     search, 608-609, 350-355. http://dx.doi.org/10.4028/www.scientific.net/AMR.608-
     609.350

10. Yetim, H., Sagdic, O. and Ozturk, I. (2008) Fatty Acid Compositions of Cold Press
     Oils of Seven Edible Plant Seeds Grown in Turkey. Chemistry of Natural Com-
     pounds, 44, 634-636. http://dx.doi.org/10.1007/s10600-008-9131-y

11. Mohanty, A.K. , Tummala, P., Liu, W., Misra, M., Mulukutla, P.V. and Drzal, L.T.
     (2005) Injection-Molded Biocomposites from Soy Protein Based Bioplastic and
     Short Industrial Hemp Fiber. Journal of Polymers and Environment, 13, 279-285.
     http://dx.doi.org/10.1007/s10924-005-4762-6

12. Berot, S., Compoint, J.P. , Larre, C., Malabat, C. and Gueguen, J. (2005) Large-Scale
     Purification of Rapeseed Proteins (Brassica napus L.). Journal of Chromatography
     B, 818, 35-42. http://dx.doi.org/10.1016/j.jchromb.2004.08.001

13. Demirbas, A.H. and Demirbas, I. (2007) Importance of Rural Bioenergy for Devel-
     oping Countries. Energy Conversion and Management, 48, 2386-2398. http://dx.doi.
     org/10.1016/j.enconman.2007.03.005

14. Rombaut, N., Savoire, R., Thomasset, B., Belliard, T., Castello, J., Hecke, E.V. and
     Lanoiselle, J. (2014) Grape Seed Oil Extraction: Interest of Supercritical Fluid Ex-
     traction and Gas-assisted Mechanical Extraction for Enhancing Polyphenol Co-Ex-
     traction in Oil. Comptes Rendus Chimie , 17, 284-292. http://dx.doi.org/10.1016/j.
     crci.2013.11.014

15. Biswas, S., Biswas P. and Kumar, A. (2012) Catalytic Cracking of Soybean Oil with
     Zirconium Complex Chemically Bonded to Alumina Support without Hydrogen.
     International Journal of Chemical Sciences and Applications, 3, 306- 313.

16. Chin, H.F. , Krishnapillay, B. and Stanwood, P.C. (1989) Seed Moisture: Recalci-
     trant vs. Orthodox Seeds. Seed Moisture, 14, 15-22.

17. Lam, P.S., Sokhansanj, S., Bi, X., Lim, C.J. , Naimi, L.J. , Hoque, M., Mani , S.,
     Womac, A.R. , Ye, X.P. and Narayan, S. (2008) Bulk Density of Wet and Dry Wheat
     Straw and Switchgrass Particles. Applied Engineering in Agriculture, 24, 351-358.
     http://dx.doi.org/10.13031/2013.24490

18. Chen , Y., Wang, C., Lu, W. and Yang, Z. (2010) Study of the Co-Deoxy-Liquefac-
     tion of Biomass and Vegetable Oil for Hydrocarbon Oil Production. Bioresource
     Technology, 101, 4600-4607. http://dx.doi.org/10.1016/j.biortech.2010.01.071

19. Rodenbush, C.M. , Hsieh, F.H. and Viswanath, D.S. (1999) Density and Viscosity
     of Vegetable Oils. Journal of the American Oil Chemists' Society, 76, 1415-1419.
     http://dx.doi.org/10.1007/s11746-999-0177-1

20. Jimenez , J.J. , Bernal, J.L. , del-Nozal, M.A., Martin , M.A. and Bernal, J. (2006)
     Sample Preparation Methods for Beeswax Characterization by Gas Chromatogra-
     phy with Flame Ionization Detection. Journal of Chromatography A, 1129, 262-272.
     http://dx.doi.org/10.1016/j.chroma.2006.06.098

21. Hamerton, I., Emsley, A.M., Hay, J.N. , Herman , H., Howlin, B.J. and Jepson, P. (2006) The Development of Controllable Complex Curing Agents for Epoxy Resins Part 3. An Investigation of the Shelf Life and Thermal Dissociation Behavior of Bis(acetanilido)-tris(acetato)dicuprate(II). Journal of Materials Chemistry, 16, 255-265. http://dx.doi.org/10.1039/b510393b

22. Requena, J.F. , Guimaraes, A.C. , Alpera, S.Q. , Gangas, E.R. , Navarro, S.H. , Gracia, L.M. , Gil , J.M. and Cuesta, H.F. (2011) Life Cycle Assessment (LCA) of the Biofuel Production Process from Sunflower Oil, Rapeseed Oil and Soybean Oil. Fuel Processing Technology, 92, 190-199. http://dx.doi.org/10.1016/j.fuproc.2010.03.004

23. Peterson , J., Carlson , R., Richter , B. and Knowles, D. (2009) Extraction of Oil from Oilseeds Using Accelerated Solvent Extraction (ASE). LCGC.

24. Laaniste, P., Joudu, J. and Eremeev, V. (2004) Oil Content of Spring Oilseed Rape Seeds According to Fertilisation. Agronomy Research, 2, 83-86.

25. Bezergianni, S., Voutetakis, S. and Kalogianni, A. (2009) Catalytic Hydrocracking of Fresh and Used Cooking Oil. Industrial & Engineering Chemistry Research, 48, 8402-8406. http://dx.doi.org/10.1021/ie900445m

26. Santillan-Jimenez, E., Morgan , T., Lacny, J., Mohapantra, S. and Crocker , M. (2013) Catalytic Deoxygenation of Tri- glycerides and Fatty Acids to Hydrocarbons over Carbon-supported Nickel. Fuel, 103, 1010-1017. http://dx.doi.org/10.1016/j. fuel.2012.08.035

27. Rustan, A.C. and Drevon, C.A. (2005) Fatty Acids: Structures and Properties. Encyclopedia of Life Sciences, 1-7.

28. Yaliwal, V.S., Daboji, S.R., Banapurmath, N.R. and Tewari, P.G. (2010) Production and Utilization of Renewable Liquid Fuel in a Single Cylinder for Stroke Direct Injection Compression Ignition Engine. International Journal of Engineering Science and Technology, 2, 5938-5948.

29. Jet A/Jet A-1 (1999) Environment Canada, Emergencies Science and Technology Division (Bulletin, Data from Shell). http://www.etc-cte.ec.gc.ca/databases/Oilproperties/pdf/WEB_Jet_A-Jet_A-1.pdf

30. Qu, W., Wei , L. and Julson, J. (2013) An Exploration of Improving the Properties of Heavy Bio-Oil. Energy & Fuels, 27, 4717-4722. http://dx.doi.org/10.1021/ ef400418p

31. Fadock, M.N. (2010) Carbon Profile Matching, Algae Fatty Acids and Jet A Fuel Properties. Guelph Engineering Journal, 3, 1-8.

32. Li, Y., Shao, J., Wang, X., Yang, H., Chen, Y., Deng, Y., Zhang, S. and Chen, H. (2013) Upgrading of Bio-Oil: Removal of the Fermentation Inhibitor (Furfural) from the Model Compounds of Bio-Oil Using Pyrolytic Char. Energy & Fuels, 27, 5975-5981. http://dx.doi.org/10.1021/ef401375q

33. Zhang, J., Toghiani, H., Mohan, D., Pittman Jr., C.U. and Toghiani, R.K. (2007) Product Analysis and Thermodynamic Simulations from the Pyrolysis of Several Biomass Feedstocks. Energy & Fuels, 21, 2373-2385. http://dx.doi.org/10.1021/ ef0606557

34. Ofori-Boateng, C., KeatTeong, L. and JitKang, L. (2012) Comparative Exergy Analyses of Jatropha curcas Oil Extraction Methods: Solvent and Mechanical Extraction Processes. Energy Conversion and Management, 55, 164-171. http://dx.doi. org/10.1016/j.enconman.2011.11.005

35. Zhang, H., Cheng, Y., Vispute, T.P. , Xiao, R. and Huber , G.W. (2011) Catalytic Conversion of Biomass-Derived Feedstocks into Olefins and Aromatics with ZSM-5: The Hydrogen to Carbon Effective Ratio. Energy & Environmental Science, 4, 2297-2307. http://dx.doi.org/10.1039/c1ee01230d

There are two tables that are not available in this version of the article. To view this additional information, please use the citation on the first page of this chapter.

PART V

CONCLUSION

CHAPTER 11

An Innovative Path to Sustainable Transportation

DANIEL SPERLING

Contrary to popular belief, the world is awash in fossil energy, much of which can be readily converted into fuels for our cars, trucks, and planes. We are not running out of fossil fuels.

The abundance of fossil fuels means we are unlikely to see high fuel prices due to scarcity. Indeed, most analysts predict that future oil prices will not be much higher than today's, apart from occasional peaks due, for example, to conflicts in the Middle East. Prices might even end up lower as new exploration and extraction technologies for shale oil, heavy oils, deep-sea oil, and oil sands make it cheaper and easier to extract fossil energy. Thus, we cannot depend on high oil prices to reduce transport energy use and greenhouse gas (GHG) emissions.

There are already 1.2 billion vehicles on the world's roads, most of which are in the rich countries of Europe, North America, and Japan. Billions more people will buy vehicles over the next century, especially in developing countries such as China and India. This raises the question: can wealthy countries not only curb their appetite for fossil energy, but also lead in developing and adopting new low-carbon lifestyles? How can mobility and accessibility be increased without disrupting ecosystems, altering the climate, depleting water supplies and extinguishing species?

We need to wean ourselves off fossil fuels—and the GHG emissions they produce—and rebuild our cities and transportation systems to be far more energy efficient. We need to shift to a world that relies on sun, water, wind, and plants for energy. We need visions, strategies, and action.

Crafting visions to reduce GHG emissions can focus public attention on what is possible and what transformations are needed. But far more important and far more challenging is the question of how we build pathways to get from here to there, with specific incremental steps to achieve desirable long-term outcomes. These pathways require a mix of strategies and technologies, which can be sorted into three legs of the transportation stool: mobility, fuel type, and vehicle efficiency. These categories are represented by a simple formula:

Mobility (vehicle miles traveled)
x Carbon Intensity of Energy (GHG emissions/unit of energy)
x Vehicle Energy Efficiency (energy use/vehicle mile)
= GHG Emissions

11.1 MOBILITY: BEYOND CAR-CENTRIC CITIES

The first leg of the transportation stool is mobility, which includes vehicle use and the infrastructure and land use that support it. In many rich countries, and increasingly in emerging economies, personal cars dominate metropolitan travel. In the US, for example, personal vehicles account for 85 percent of passenger miles traveled, air travel for about 10 percent, public transportation for less than 3 percent, and bicycles and walking less than 2 percent. In Europe, where cities are more compact and fuel prices much higher, public transport has shrunk to only about 16 percent of passenger miles traveled. Walking accounts for about 20 percent of trips in much of Europe and Japan, though the trips are short and thus account for only a small share of total travel.

Reliance on personal vehicles makes transforming our cities and transportation systems a daunting challenge. In the US, passenger transport has changed little in over 60 years. While cars are safer, more reliable, and more energy efficient, they continue to perform at similar speeds and capacities and continue to be powered by internal combustion engines and petroleum. Highways, too, have changed little over the last 60 years, and for the most part remain toll-free, serve all vehicle types, and provide roughly the same performance.

Transit, too, has seen little innovation since the 1950s. The concept of bus rapid transit (BRT), where buses operate in platoons over dedicated rights-of-way, is one of the few significant innovations in intervening decades. Aside from BRT, transit operates with nearly identical service, capacity, and performance. US rail transit technology has improved, but rail ridership has decreased; heavy and light rail systems together carry less than one percent of passenger travel. Even in Europe, where ridership is more robust, rail transit carries only seven percent of overall passenger travel.

We need much more innovation. We need a new model of mobility and accessibility, especially in rapidly expanding economies such as Brazil, China, and India. Rich countries are now largely locked into car-centric development, but even car-centric cities still have the opportunity to reduce car use by expanding innovative mobility services, improving communication technologies, and charging full-cost prices for vehicles, fuels, roads, and parking. Better land use policies, too, will reduce vehicle use.

One promising legislative initiative is California's "Sustainable Communities Act of 2008," known as SB375. SB375 imposes on cities a greenhouse gas target for passenger travel and provides a framework to transform the mix of mobility services, land use, institutions, and behaviors that underlie our cities.

The information and communication revolution is bringing efficient, cost effective mobility services that can fill the gap between single-occupant cars and bus and rail transit. Vans and small buses can respond to real-time trip requests through demand-responsive transit services. Travelers located near each other self-organize carpools in real time using smart carpooling. And smart car and bike sharing facilitate one-way trips through shared vehicle use. Proliferating start-up companies are beginning to offer all these services, but they still account for only a tiny share of urban travel.

Additional benefits can come from enhanced data gathering and road pricing to improve traffic management. Innovations such as high occupancy toll lanes are a step toward more rational use of roads, and collection of "big data" from sensors and smart phones will lead to better design and management of roads, public transit, and new mobility services. Another innovation—automated vehicle technology—provides more potential for greater access at lower cost (but also runs the risk of encouraging more driving).

These strategies are not original ideas, and many have not yet been vetted or embraced. Concerns about privacy, opposition from taxi companies, and liability concerns all threaten the spread of innovative mobility services. But changes are afoot. Many new companies provide real-time mobility services, and some are preparing to market small neighborhood cars. Some cities use pricing to manage traffic flows and parking. Other cities are establishing bus rapid transit systems.

Unfortunately, these initiatives are fragmented in scattered cities. Therein lies the real challenge. No one technology, service, or land use change will substantially reduce car dependency by itself. Travelers will use more sustainable modes in large numbers only if they gain access to a suite of mobility services, reinforced with pricing and better non-car infrastructure.

11.2 REDUCING THE CARBON INTENSITY OF FUELS

Vehicle fuel, the second leg of the transportation stool, adds carbon and other greenhouse gases to our atmosphere and threatens our climate. Oil companies are investing hundreds of billions of dollars in unconventional oil production such as shale oil, heavy oils, deep-sea oil, and oil sands.

They focus on carbon-intense unconventional oil production, rather than renewable energy, for two reasons: 1) OPEC countries have nationalized their oil supplies, thereby reducing access by large investor-owned western oil companies to the most abundant conventional oil reserves; and 2) oil companies' core competencies are best suited to building large (fossil) projects rather than small renewable energy projects.

To invest in a sustainable future, we must shift investment from fossil sources to renewable sources, but replacing petroleum will be difficult and slow. The hegemony of petroleum creates multiple barriers for new fuels including liability, public skepticism, and sunk investment in supportive infrastructure like refineries, pipelines, and fuel stations. Politicians and the media are quick to embrace new transportation fuels—methanol in the 1980s, electric vehicles in the early 1990s, and hydrogen in the early 2000s—but the barriers and the inertia inevitably hamper newer fuels' potential. This fuel du jour phenomenon, whereby new fuels are hyped and then quickly abandoned, describes the recurring failures of alternative fuels over the past few decades.

Looking forward, there are three sets of promising energy options. The first is electricity for vehicles, which provide large environmental benefits, and battery technology is steadily improving. But will the short attention span of politicians and media again kill off this attractive alternative before it has time to gain consumer acceptance, achieve scale economies, benefit from learning-by-doing, and fully realize the benefits of supportive policies?

Hydrogen is a second promising option. Hydrogen may have the greatest potential to replace petroleum across transportation, but it faces even greater start-up and fuel-distribution challenges. Not only must the vehicle industry transform from combustion engines to fuel cells, but the fuel industry must also shift from supplying gasoline and diesel fuel to supplying hydrogen.

Biofuels are the third major contender to reduce oil dependency. A downside to biofuels is that they require ample land. Diverting land to energy production is problematic because it reduces farming areas and releases huge amounts of carbon embedded in soils and plants. Biofuels are most promising if made from urban, forestry, and crop wastes, which do not require additional land.

Policies should reward and support options that are most likely to re-duce oil dependency and carbon intensity. Performance-based policies that reward carbon reduction, such as the low carbon fuel standard in Califor-nia, are promising, especially when accompanied by incentives that help boost initial investments in hydrogen stations and advanced biofuel tech-nologies. Without strong policies, high-GHG petroleum fuels will sweep aside all alternatives, including electric vehicles.

11.3 ROLLING OUT EFFICIENT VEHICLES

Improving vehicle efficiency is the third and most promising strategy for reducing GHG emissions for the next few decades. Aggressive policies in most large car markets, including Europe, the US, China, and Japan, aim to cut fuel consumption per vehicle roughly in half in the next 15 years. Importantly, the international auto industry has made energy efficiency a top priority. Car manufacturers are using more lightweight materials, more efficient transmissions, better combustion technology, improved aerody-namics, and hybrid engines. They are also developing pure battery-electric vehicles, plug-in electric hybrids, and hydrogen-powered fuel cell elec-tric vehicles. By 2035, a large proportion of new vehicle sales around the world are likely to be plug-in and fuel cell vehicles. The combination of efficiency improvements and electrification could lead to an 80 percent reduction in greenhouse-gas emissions for light duty vehicles (per vehicle-mile) by 2050.

Vehicle electrification is key to the long-term sustainability of vehicles. Electrification includes a spectrum of technologies, from those operating on gasoline or diesel fuel but assisted by batteries and electric motors, to those operating solely on electricity and/or hydrogen.

Today, every major car company is actively pursuing zero-emission technologies, and most have such vehicles in production. The challenge is to sustain the momentum. Costs of new technology are always high. Strong, durable policies are needed to keep automakers and consumers engaged, including the following:

- Incentives and mandates for early vehicle production,

- Government support of charging and hydrogen supply infrastructure,
- Market-based policies that internalize the cost of climate change and provide incentives to consumers and automakers to buy and sell low-carbon vehicles and fuels, and
- Performance-based standards to provide a regulatory framework for automakers as they plan for the future.

While cars are a big success story, trucks are not. Truck efficiency improvements are much slower and early regulations in Europe, the US, and Japan are much weaker than those for light duty vehicles. More stringent performance standards are likely to be developed in the future. The bigger obstacle, though, is that most current low-carbon energy alternatives are unsuitable for trucks. Trucks are heavy and tend to travel long distances each day. Most trucks cannot be easily powered by electricity because the batteries needed to provide sufficient energy are very heavy and bulky. Fuel cells may be a more feasible option for trucks, since they are far lighter and smaller than batteries. The best long-term energy option for trucks, however, will probably be low-carbon biofuels.

11.4 CONCLUSION

We need to change how we harness and use energy for transport. All of us need to confront the reality that transportation as we know it is incredibly expensive, resource-intensive, and socially unjust. We need to improve our technologies and learn new behaviors. But new technologies and new behaviors do not appear spontaneously; they take time. Climate change, a compelling reason for these new technologies and behaviors, is not directly experienced, unlike the more observable problems of smog, polluted water, and marred landscapes. With no direct, observable consequences to increased greenhouse gas emissions, it is difficult to pursue new transportation and energy pathways that are initially more expensive and less convenient.

And yet, sweeping changes are needed. If we don't make those tough economic choices and major lifestyle changes, then we condemn our grandchildren and great-grandchildren to an environmentally compromised world. Who has the courage to step forward and lead? It is morally and economically irresponsible to leave this looming disaster as our legacy. We need to muster the necessary courage and channel our innovative spirit if we hope to create a better world.

FURTHER READINGS

1.   David L. Greene, Sangsoo Park, and Changzheng Liu. 2014. "Analyzing the Transition to Electric Drive Vehicles in the U.S.," Futures, 58: 34–52.
2.   National Research Council. 2013. Transitions to Alternative Vehicles and Fuels, Washington, DC: The National Academies Press.
3.   Susan Shaheen and Adam P. Cohen. 2013. "Carsharing and Personal Vehicle Services: Worldwide Market Developments and Emerging Trends," International Journal of Sustainable Transportation, 7(1): 5–34.
4.   Daniel Sperling and Deborah Gordon. 2009. Two Billion Cars, Oxford, UK: Oxford University Press, (paperback, 2010).
5.   Daniel Sperling and Mary Nichols. 2012. "California's Pioneering Transportation Strategy," Issues in Science and Technology, Winter: 59–66.

This article is adapted from the Commemorative Lecture that Daniel Sperling delivered in Tokyo when he was awarded the Blue Planet Prize in 2013. http://www.af-info.or.jp/en/blueplanet/doc/prof/2013profile-eng.pdf

Author Notes

CHAPTER 1

Acknowledgments
The Authors wish to thank the FSE – European Structural Funds POSDRU with the Financing Contract POSDRU/107/ 1.5/S/76909 (ValueDoc), for the support offered during the research.

CHAPTER 2

Acknowledgments
The authors would like to thank Neal Carboneau and John Habermann for their useful comments. The contents of this paper reflect the views of the authors who are responsible for the facts and the accuracy of the information presented herein and do not necessarily reflect the official views or policies of the FHWA and INDOT nor do they constitute a standard, specification, or regulation.

Competing Interests
The authors declare that they have no competing interests.

Author Contributions
KAW and WR collected the data and conducted the analysis. PA and KAW drafted the manuscript. JH, JF, and PA led and coordinated the study and affected its design. All authors read and approved the final manuscript.

CHAPTER 3

Acknowledgments and Disclaimer
This work is funded by the California Department of Transportation, Division of Research, Innovation and System Information, and the University

of California Institute of Transportation Studies, Multi-campus Research Programs and Initiatives. The California LCA project is part of a pooled-effort program with eight European national road laboratories and the Federal Highway Administration called Models for rolling resistance In Road Infrastructure Asset Management Systems (MIRIAM).

The opinions and conclusions expressed in this letter are those of the authors and do not necessarily represent those of the California Department of Transportation, the Federal Highway Administration, the University of California, or the MIRIAM consortium.

CHAPTER 4

Acknowledgments
This project was funded by California Energy Commission's Public Interest Energy Research Program under contract 500-10-009. The authors would like to thank Emmanuel Liban and other staff at LA Metro, Alberto Ayala and Shaohua Hu (California Air Resources Board), Paul Bunje (UCLA), Julia Campbell (UCLA and LA Metro), Pierre DuVair (California Energy Commission), and Andrew Fraser (Arizona State University) for their support and input.

CHAPTER 5

Acknowledgments
We acknowledge the Gordon and Betty Moore Foundation for funding. We thank the Metropolitan Transportation Commission and the Bay Area Congestion Management Agencies for their support in data collection. We thank Paul Medved (BART) for providing information about rail transportation development and impacts.

CHAPTER 6

Acknowledgments
This work was developed under STAR Fellowship Assistance Agreement no. FP917275 awarded by the US Environmental Protection Agency

(EPA). It has not been formally reviewed by EPA. The views expressed are solely those of the authors, and EPA does not endorse any products or commercial services mentioned.

CHAPTER 7

Acknowledgments
The authors would like to acknowledge the Engineering and Physical Sciences Research Council (EPSRC) who fund the 'Airports and Behavioural Change (ABC): towards environmental surface access' project (EP/H003398/1). They would also like to thank staff at the two airports, Manchester Airport and Robin Hood Doncaster Sheffield for their support and advice during the research.

Conflict of Interest
The authors declare no conflict of interest.

CHAPTER 9

Competing Interests
The authors declare that they have no competing interests.

Author Contribution
EH carried out the whole manuscript despite of the Technology section, which was carried out by both authors. PM carried out the electric conversion of the used Smart. EH corrected and approved the final manuscript.

Acknowledgments
The authors like to thank Katharina Schowalter, Rüdiger Hild and Bernd Fuss for data provision and Viola Helmers for the linguistic comments on the manuscript. We also appreciate the advice by M. Gauch and particularly G. Majeau-Bettez.

CHAPTER 10

Acknowledgments
This study was funded by the U.S. Department of Transportation through NC Sun Grant Initiative under Grant No. SA0700149. The authors would like to thank the Chemical Analytic Lab in the Chemistry Department of South Dakota State University for the GC-MS analysis of the oil samples. All the support is gratefully acknow- ledged. However, only the authors are responsible for the opinions expressed in this paper and for any possible error.

Index

T - #0829 - 101024 - C320 - 229/152/14 - PB - 9781774636978 - Gloss Lamination